PRO

Promise Theory bridges the worlds of semantics and dynamics to describe scalable interactions between autonomous agents that form clusters and groups. It provides a broadly developed and semi-formal language, which builds on the mathematics of sets and graphs, and models intent and outcome in an impartial manner. The result is a theory that expresses a 'chemistry' of cooperative behaviours for a wide range of systems, emphasizing how each new scale of cooperation leads to new phenomena and new promises.

This book is aimed at scientists, philosophers, and engineers. It introduces readers to the key concepts in a practical manner, building on the foundation of voluntary cooperation as a ground state for all interacting systems. The book draws on many examples from the real world, with a particular emphasis on human-computer systems.

'*Promise Theory offers a methodology for generating certainty on top of uncertain foundations. This book presents the formal foundations of Promise Theory. It lays out the formalisms in a clear, concise, understandable way that makes them accessible to non-mathematicians. If you want to fully understand the conceptual mechanisms that underlie the distributed systems that make up today's "cloud services", you should start with this book.*'
– Jeff Sussna, Author of Designing Delivery

'*[The authors] bring the rigor of theoretical physics to the science of cooperation. The application of this kind of rigor to the social sciences is a tremendous leap forward. [The] pioneering work on developing an algebra of cooperation is an idea whose time has come. A promise is not a guarantee. That said: I promise you that examining this book will stimulate your thinking about cooperation and collaboration at scale. This book covers a lot of ground: promises, impositions, invitations, games, and the peculiar dynamics of authority and authorization. Those looking for a book that applies the lessons of distributed computing to the new and emerging science of cooperation will find what they are looking for here.*'
– Daniel Mezick, Author of The Culture Game and Inviting Leadership

Also by the authors:

PROMISE THEORY: CASE STUDY ON THE 2016 BREXIT VOTE

MONEY, OWNERSHIP, AND AGENCY—AS AN APPLICATION OF PROMISE THEORY

Other reviews:

'*A landmark book in the development of our craft...*'
–Adrian Cockcroft (about *In Search of Certainty*)

'*Proud to say that I am a card-carrying member of the [Mark Burgess] fan club. And I think that it wouldn't be too much of a stretch to say that he's the closest thing to Richard Feynman within our industry (and not just because of his IQ).*' –Cameron Haight (about *Smart Spacetime*)

'*...our whole industry is transforming based on ideas [Mark Burgess] pioneered*'
–Michael Nygard (about *Smart Spacetime*)

'*The work done by [Mark] on complexity of systems is a cornerstone in design of large scale distributed systems...*'
–Jan Wiersma (about *In Search of Certainty*)

'*Some authors tread well worn paths in comfortable realms. Mark not only blazes new trails, but does so in undiscovered countries.*'
–Dan Klein (about *Smart Spacetime*)

PROMISE THEORY

PRINCIPLES AND APPLICATIONS
SECOND EDITION

JAN A. BERGSTRA AND MARK BURGESS

χtAxis press

First published by χtAxis press 2014.
Reprinted with corrections, October 2014, August 2017.
Second edition published 2019, under the imprint χtAxis press

Text and figures Copyright © Jan A. Bergstra and Mark Burgess 2004-2019.

Jan Bergstra and Mark Burgess have asserted their right under the Copyright, Design and Patents Act, 1988, UK, to be identified as the authors of this work.

All rights reserved. No part of this publication may be copied or reproduced in any form, without prior permission from the author.

Cover design by Zhaoling Xu.

Contents

1 What Does Promising Mean?   1
    1.1   A practical theory of promises   1
    1.2   Autonomous agents make promises   2
    1.3   Tenets of Promise Theory   3
    1.4   The main concepts   4
    1.5   'Many worlds' and different observers   4
    1.6   No special status for humans   6
    1.7   The simple essence of intent   7
    1.8   The history of promise concepts   8
    1.9   Promises in information technology   8
    1.10   Promises and trust are symbiotic   8
    1.11   How to use Promise Theory   9

I Fundamentals   11

2 The Heuristics of Promise Theory   13
    2.1   What qualifies as a promise?   13
    2.2   Promises by inanimate agents   14
    2.3   A model of a promise   15
    2.4   Impositions   17
    2.5   Implicit promise versus explicit promise   17
    2.6   The role of promises to reduce uncertainty   18
    2.7   Obligations   18
        2.7.1   Meaning of obligations   19
        2.7.2   Scope of obligations   19
        2.7.3   Obligations and force as possible intentions   19
    2.8   For and against the primacy of obligations   20

	2.8.1 In favour of obligations	20
	2.8.2 Against obligations	20
2.9	Assessing promises	22
2.10	The implication of assessment	23
2.11	The value of promises	23
2.12	Deceptions as promises	24
2.13	More elements of Promise Theory	25

3 Formalizing Promises 26

3.1	Agents	26
3.2	Formal (micro)promises defined	26
3.3	Promise proposals and signing	28
	3.3.1 What we can say about the body?	28
3.4	Notation and conventions	29
3.5	Scope, worlds, and relativity	30
	3.5.1 Denoting information about promises	30
	3.5.2 Denoting promise scope	31
	3.5.3 Types and labels	31
3.6	Impositions	32
3.7	Obligations	33
3.8	Kinds of promises	33
	3.8.1 Promises of the first kind	34
	3.8.2 Promises of the second kind	34
	3.8.3 Promises of the third kind	34
	3.8.4 Promises of the fourth and most general kind	35
	3.8.5 Elementarity of promises of the first kind	36
3.9	Promise graphs and networks	36
3.10	Defining promise matrices	37
3.11	Transfer of intentional behaviour between agents	38
	3.11.1 Polarity $\pm$ of promises	38
	3.11.2 Promises to use	38
	3.11.3 The use promise as a non-primitive	39
	3.11.4 Polarity of $\pm$ impositions	39
	3.11.5 Cooperative interactions between agents	40
	3.11.6 Complementarity and $\pm$ duality: push vs pull	41
	3.11.7 The effective action of a cooperative binding	43
3.12	Particular kinds of promises	43
	3.12.1 Commitments	43

		3.12.2	Exact and inexact promises	44
		3.12.3	Superfluous or empty promises	44
		3.12.4	Coordination promises	45
		3.12.5	Lies and deceptions	45
		3.12.6	Promises to keep promises	46
	3.13	Idempotence in repetition of promises		47
	3.14	Promises involving groups, anonymous and wildcard agents		48
	3.15	Goals or ranking of intentions		49

4 Combining Promises into Patterns — **50**

	4.1	Bundles of promises		50
	4.2	Parameterized bundles		51
	4.3	Promise valency and saturation		52
	4.4	Agent roles		54
		4.4.1	Formal definitions of kinds of role	54
		4.4.2	Graphical interpretation of roles.	57
		4.4.3	Symmetry of agents under roles	57
	4.5	Some common patterns or roles		58
		4.5.1	Client or customer	58
		4.5.2	Consumer or dependant	58
		4.5.3	Server or supplier	59
		4.5.4	Server pool (coordinated servers)	59
		4.5.5	Peer (vice-versa exchange)	59
		4.5.6	Queue dispatcher or forwarding agent	60
		4.5.7	Transducer	61
		4.5.8	The matroid hub construction (calibration)	62

5 Assessment and Measurement — **65**

	5.1	Elements of promises and assessments		65
	5.2	Defining assessment		66
		5.2.1	A definition of promise assessment	66
		5.2.2	Auxiliary definitions of assessment	67
		5.2.3	True, false and indeterminate	68
		5.2.4	Inferred promises: emergent behaviour	68
	5.3	Boundary conditions for assessments		69
	5.4	The observation interaction: measurement		69
	5.5	Algebra of measurement		71
		5.5.1	Distinguishability and equivalence of outcomes	72
		5.5.2	Rewriting for cooperative behaviour	74

	5.6	Emergent behaviour defined	75
6	**Promise Mechanics**		**76**
	6.1	The laws of promise composition	76
		6.1.1 Multiple promises by a single agent	76
		6.1.2 Ensemble promises	77
		6.1.3 Effective action and coupling strength during transfer of intentional behaviour	77
	6.2	The Laws of Conditional Promising	78
		6.2.1 Promises with conditions attached	78
		6.2.2 Complementarity of promises with conditionals	80
	6.3	Modelling cooperative behaviour	81
		6.3.1 Subordination and autonomy	83
7	**Promise Dynamics**		**85**
	7.1	The promise life-cycle	85
	7.2	Agent worlds and knowledge dissemination	85
		7.2.1 Definition of knowledge	86
		7.2.2 Distributed consistency of knowledge	87
		7.2.3 Assimilation of knowledge	89
		7.2.4 Integrity of information through intermediaries	89
		7.2.5 Uniformity and common knowledge	90
	7.3	Laws of behaviour for agents	91
	7.4	Deadlock in conditional promises	92
		7.4.1 From equilibrium to an arrow of time	93
8	**Reasoning About Promises**		**95**
	8.1	Exclusive or incompatible promises	95
	8.2	Promise conflicts	96
		8.2.1 Breaking promises	99
		8.2.2 Unintended promise breaking	99
		8.2.3 Indirection and conflicts	100
	8.3	Promise refinement and overriding	100
	8.4	Agreement	102
		8.4.1 Agreement as promises	102
		8.4.2 Contractual agreement	102
		8.4.3 Contracts and signing	103
		8.4.4 Cooperative agreement or treaty	103

9	**The Value of Promises**	**104**
	9.1 Value function of a promise	104
	9.2 Valuation function	105
	9.3 Mean Time to Keep a Promise	106
	9.4 Some more possible valuations	106

10	**Trust and Promise Keeping**	**108**
	10.1 Trust and autonomous promises	108
	10.2 A compatible notation for trust	109
	10.3 How promises guide expectation	111

II Applications   113

11	**Workflows and End-To-End Delivery**	**115**
	11.1 Intermediaries or proxies	115
	11.2 Simple process models in human systems	116
	11.2.1 Buying milk	116
	11.2.2 Booking a doctor appointment	117
	11.2.3 Hiring a contractor or employee	118
	11.2.4 Military systems	119
	11.3 Promising state versus promising change	119
	11.4 Delivery chains	119
	11.4.1 Relative delta-push delivery pattern	120
	11.4.2 Delivery pattern for one intermediary	120
	11.4.3 Desired end-state delivery pattern for multiple intermediaries	122
	11.4.4 End-to-end integrity	125
	11.5 Transformation chains or assembly lines	125

12	**Transport Network Systems**	**127**
	12.1 Virtual circuits	127
	12.2 Application services	129
	12.3 Multi-component promise architectures	129
	12.3.1 Case (a): fragile load balancing with capacity bottleneck	130
	12.3.2 Case (b): fragile load balancing with queue bottleneck	132
	12.3.3 Case (c): robust load balancing with no bottleneck	133
	12.3.4 Remarks about systemic promises	133

13 Systemic Promises   135

- 13.1 Agent type differentiation 135
  - 13.1.1 Organization of agents into ordered states 136
  - 13.1.2 Specialization and functional reusability 137
  - 13.1.3 Organizations or institutions 137
- 13.2 Superagents: agents made of other agents 138
  - 13.2.1 Composition of agents (sub and superagency) 139
  - 13.2.2 superagent surface boundary 142
  - 13.2.3 Irreducible promises at scale M, and collective behaviour . . . 143
- 13.3 Centralization, decentralization, and hierarchy 146
- 13.4 Occupancy and tenancy of space 148
  - 13.4.1 Definitions of occupancy and tenancy 148
  - 13.4.2 Laws of tenancy semantics 150
  - 13.4.3 Forms of tenancy 153
  - 13.4.4 Tenancy and conditional promises 156
  - 13.4.5 Remote tenancy 156
  - 13.4.6 Asymmetric tenancy 157
  - 13.4.7 Scaling of occupancy and tenancy 158
  - 13.4.8 Distribution or dispatch of promises at superagent boundaries . 159
  - 13.4.9 Adjacency between external agents and superagents 161
- 13.5 Human systems – 'soft topics' 163
  - 13.5.1 Seeking cooperation and support of other agents 163
  - 13.5.2 Invitation, attack, and command 165
  - 13.5.3 A more refined view of invitation vs attack 166
  - 13.5.4 The intent to disrupt or sabotage 173
  - 13.5.5 Accusations, the imposition of judgement, and taking offense 174
- 13.6 Responsibility 175
  - 13.6.1 Subjectivity in assessments 175
  - 13.6.2 The role of conditional promises in pointing to responsibility 176
  - 13.6.3 Downstream principle 178
  - 13.6.4 Assuming responsibility 180
- 13.7 Rights, permission, and privileges 181
  - 13.7.1 Rights defined 181
  - 13.7.2 Seeking rights and permissions 183

	13.8	Authority, power, and delegation	184
	13.9	Trusted Third Parties and Webs of Trust	187
	13.10	Leadership in a human system	190

14 Componentization and modularity in systems — 192

	14.1	Definition of components	193
	14.2	What systemic promises should components keep?	193
	14.3	Component design and roles	194
	14.4	Promise conflicts in component design	194
	14.5	Reusability of components	196
		14.5.1 Definition of reusability	196
		14.5.2 Interchangeability of components	197
		14.5.3 Compatibility of components	197
		14.5.4 Backwards compatibility and regression testing	198
		14.5.5 Generic interface expectations	199
		14.5.6 Irrevocable component choices	201
		14.5.7 Versioning, evolution and promise conflicts	202
		14.5.8 Naming and 'branding' of components	204
		14.5.9 Naming component usage or promise configurations	205
	14.6	The economics of components	206
		14.6.1 The costs of modularity	206
		14.6.2 Addendum to choosing between alternative components: fitness for purpose	208

15 Promises and Mathematical Games — 209

	15.1	Relationship between games and promises	209
	15.2	Classification of games	211
		15.2.1 Cooperative, no conflict	211
		15.2.2 Cooperative, with conflict	212
		15.2.3 Non-cooperative, no conflict	212
		15.2.4 Non-cooperative, with conflict	212
		15.2.5 Constant-sum games	213
		15.2.6 Multiple strategies and mixtures	215
	15.3	Repeated bargaining: conditional promises	216
	15.4	Promises: constructing policy	219
	15.5	Management dilemmas	220
	15.6	Examples of games and the economics of promises	222
	15.7	Minimum incentive requirements	225

16 Semantic Spacetime — 227
- 16.1 Extended agent structure 227
  - 16.1.1 Continuity and basis 228
  - 16.1.2 Scalar promises—material properties 230
  - 16.1.3 Vector promises and quasi-transitivity 231
- 16.2 Adjacency between agents 232
  - 16.2.1 Locality 232
  - 16.2.2 Non-local extended causal order 233
  - 16.2.3 Ordered agents 235
  - 16.2.4 The meaning of a link 237
  - 16.2.5 Relationship to the end-to-end problem 238
  - 16.2.6 Co-dependence (entanglement) 240
  - 16.2.7 Boundaries and holes 241
  - 16.2.8 Containment within regions 243
- 16.3 Symmetry of short and long range order 244
- 16.4 Motion in agent space 245
  - 16.4.1 Motion of the first kind 246
  - 16.4.2 Motion of the second kind 248
  - 16.4.3 Motion of the third kind 249

17 CFEngine: A Promise keeping engine — 250
- 17.1 Policy with autonomy 251
- 17.2 History 251
- 17.3 A language of configuration promises 253
- 17.4 The software agents 254
- 17.5 The syntax of promises in CFEngine 255
- 17.6 How CFEngine keeps promises 257
  - 17.6.1 Who actually keeps the promise? 259
  - 17.6.2 How long do promises last? 260
  - 17.6.3 Coordination 260
  - 17.6.4 Promise bundles 261
- 17.7 CFEngine's design components 262
  - 17.7.1 Design guidelines for components 264
  - 17.7.2 An example format for components 264
  - 17.7.3 Conditional promises and stacking of components 265
  - 17.7.4 Composition of components 267
  - 17.7.5 Accurate naming of sketch behaviour 268

Preface to Second Edition

In this second edition, we have tried to make as few changes to the original text as possible, so that it continues to serve as a simple reference, as well as a point of historical documentation. Nevertheless, much work has been done over the years on specific details and applications. Apart from a few technical changes to notation, we've added to the applications of promises especially in human systems, with issues like authority, rights, tenancy, and the meaning of invitation, inspired by the work of others in applying promise ideas. A whole new chapter has been added about Semantic Spacetime, which has developed quickly into a viable approach to problems in physics and computer science, especially Artificial Intelligence.

Promise Theory is now fifteen years old and, over the years of studying and applying it to numerous cases, we've changed our minds about a few conventions, and developed many new concepts on the back of Promises. Luckily, the basic definitions and concepts have proven surprisingly resilient. We've also used the opportunity to improve the index coverage.

Additional volumes are now being added to the Promise Theory series, applying it specifically in particular cases, in the hope that these may inspire others to develop the picture further. This seems to be the correct approach, given the diversity of cultures and subject matter that can be addressed with the promise concept. We hope readers might seek out some of those works, and that this new edition will serve its purpose for the next phase of Promise Theory's development.

JB and MB, 2019

Preface to First Edition

Coordinating activity in an uncertain world requires flexible communication between autonomous players. As humans, we wrestle daily with the running of companies, economies, computer systems, and even whole societies. Ambiguities abound, and context-dependent meanings are both a feature and a tool in these scenarios. It is here that the idea of promises seems to be indispensable.

In this book, we explore the expressive power of promises in a variety of circumstances, challenging readers to explore this fascinating theme for themselves rather than presuming to offer all the answers. We found no satisfactory logic of promises; promises are about 'best effort' rather than the more familiar determinism of classical science. Nevertheless, we exchange promises every day, and so do artificial agents and systems, all to good effect. We focus on application rather than on justification, and readers may refer to [BB13] and [BB14] for further details of the latter.

We hope readers will share our point of view that promises deserve a prominent role in a modern story of cooperative systems.

    Jan A. Bergstra
    Mark Burgess

Chapter 1

What Does Promising Mean?

This book is about the theory of *promises* — an unusual topic, by any account. For something that so permeates daily life, the subject of promises seems to be curiously hidden away in academic literature. A thorough search reveals specialized references in areas like philosophy, economics, law, and computer science, but these hardly make up for the role played by promises in our world. Indeed, as a way of unifying interactions all the way from the microscopic world of physics to the macroscopic scale of societies, the generalized notion of promises has revealed many important insights.

Why should promises be so undersold? One possible reason for this is that the notion of promises is surprisingly difficult to formalize, requiring an even mixture of semantic and dynamic components. Another explanation, however, is that the world has been much more focused on the law-giving paradigm of *obligation*. In literature, promises are very often waived completely in favour of obligations. We imagine imposing our will, rather than seeing cooperation as a voluntary act. Some authors claim that a promise merely implies an obligation to keep the promise, and therefore the notion of a promise is redundant. This view is far from the case, and more importantly it is unnecessary. We shall show that promises are a fully independent concept, and one that brings great insight to a wide range of phenomena.

1.1 A Practical Theory of Promises

This book considers how promises may be used as a tool for understanding world around us. Promises play all kinds of roles in systems from simple mechanisms all the way up to societies. They can be casual remarks, barely taken seriously, or they can serve as technical building-blocks that characterize the structure and behaviour of a device. In some cases, promises might be complex litanies of intent, steeped in deception. In this book, we give a flavour of these different roles, but focus on how to use them, from a practical engineering viewpoint for the most part.

Promises are especially important in a world occupied by humans, alongside machines, plants, animals, and other entities[1]. Without humans, or some other cognitive agents, there seems to be little need for promises or intentions; one should be able eliminate intent and reduce behaviour to the purely mechanistic language of physics. This is naive, however. Much of natural science can also be cast in terms of a generalized notion of promises, i.e. autonomously given declarations of expected behaviour which stem from some identifiable characteristics. The aims of promise theory include[2]:

- To clarify the role of intent and its possible interpretations, within a framework that unifies dynamic and semantic aspects of behaviour.
- To define a practical notation for documenting intended and unintended behaviour.
- To derive a practical calculus or algebra for answering questions about systems of promises, e.g.
    - When are there too many or too few promises to conclude that something is likely?
    - When are the promises inconsistent with one another?

1.2 AUTONOMOUS AGENTS MAKE PROMISES

The active entities in Promise Theory are called agents. Agents can be persons, animals, plants, machines, or any other kind of entity. They are the things with the agency to exhibit behaviours, whether intentionally or unintentionally, and whose observation leads to perception of behaviours and intentions in other agents. Some of these agents can make promises through what we would call free will; others merely seem to keep effective promises through the agency of inanimate tools, e.g. a light bulb that promises to shine with a certain brightness.

Promises express the 'intended behaviour' in the broadest sense. This includes inanimate properties and attributes of agents that play active roles in human plans[Sea83]. It is *observers* who have the last word about interpreting these promises, and who decide whether the assertions are actionable, helpful, good, bad, and so on.

Autonomous implies the assumption that agents cannot be coerced into making promises, hence no agent may make promises on behalf of another. Each agent lives in its own 'private world', with access to its own information, and perhaps to information is promised by other agents, assuming that they promise to pay attention to such information.

This means they always have *incomplete information* about the world in which they live. It places limitations on what agents can know, and how they can reason about the world.

1.3 Tenets of Promise Theory

The basic tenets of Promise Theory may thus be summarized as follows. We explain these in more detail in the following chapters.

1. Agents are autonomous. They can only make promises about their own behaviour. No other agent can impose a promise upon them.

2. Making a promise involves passing information to an observer, but not necessarily a message in the explicit sense of a linguistic communication[3].

3. The assessment of whether a promise is kept or not kept may be made independently by any agent in its scope.

4. The interpretation of a promise's intent may be made independently by any agent in its scope.

5. The internal workings of agents are assumed to be unknown. Knowledge of them may be assessed from the promises they make, and keep. However, we may choose the boundary of an agent wherever we please to hide or expose different levels of information, e.g. we may think of a car as an atomic vehicle, or as a collection of agents working together.

Some level of common knowledge or common understanding (at minimum, a lingua franca for communication of basic intent) may be needed to get from autonomous agents to coordinate and promise cooperative behaviour. Without this, we cannot guarantee that all agents will understand one another's intentions [Bur15].

Before proceeding, we should like to issue a warning to readers fascinated by autonomy, by remarking on a point that has been misunderstood in the past. Promise Theory's focus on autonomous agents is by no means to be taken as an ideological position that advocates decentralization, or self-reliance in systems, no matter how appealing it might sound to some readers. Rather, it is about using that bare bones starting point to clarify precisely how effective or ineffective the strategies of centralization and decentralization can be under different circumstances. The scaling of robust intentional behaviour is a highly complicated matter that goes beyond the scope of this book, but no one should fool themselves into believing that there are simple prescriptions that will apply to all circumstances. Readers are referred to [Bur19a] for further discussion on these matters.

1.4 The main concepts

We shall refer to the following key concepts repeatedly:

- *Intention*: A subject or type of possible behaviour. i.e. something that can be interpreted to have significance. Any agent can harbour intentions. It could be something like 'be red' for a light, or 'win the race' for a sports person.

- *Promise*: When an intention is publicly declared to some audience (called its scope) it becomes a promise. Thus a promise is a stated intention.

- *Imposition*: This is an attempt to induce cooperation in another agent, i.e. to implant an intention, without prior invitation of promise to accept. It is complementary to the idea of a promise, and involves an element of the unexpected. Degrees of imposition include: hints, advice, suggestions, requests, commands, etc.

- *Obligation*: An imposition that implies a cost or penalty for non-compliance. It is more 'aggressive' than a mere imposition.

- *Assessment*: A decision about whether a promise has been kept or not. Every agent makes its own assessment about promises it is aware of. Often assessment involves the observation of other agents' behaviours.

There are other levels of interaction between agents. One could, for example speak of an attempt to force an agent to comply with an imposition, which might be termed an attack; however, we shall not discuss this further as it leads into discussions of morality which we aim to avoid.

Perhaps surprisingly, promises are more common than impositions and hence take precedence as the instrument of primary focus. Impositions generally only work at all in a system of pre-existing promises. For example, a billiard ball strike only works because the balls are elastic bodies made of atoms that repel one another electrostatically. Promises can often be posited to replace impositions with equivalent voluntary behaviours. As we shall see, this is a very useful analytical technique.

1.5 'Many worlds' and different observers

Promises and impositions are always seen from the vantage point of autonomous agents (see figure 1.1). We may call such a subjective view the 'world' of the agent, and Promise Theory is a relativistic theory of 'many worlds' belonging to its many agents.

1.5. 'MANY WORLDS' AND DIFFERENT OBSERVERS

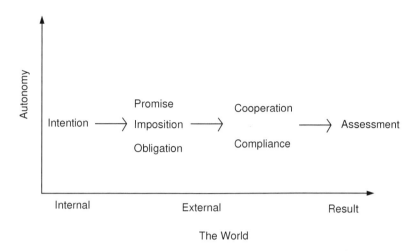

Figure 1.1: A progression of related promise concepts from left to right starts with forming an intention internally, then bringing this into the world as a promise, imposition or obligation. This leads to some degree of cooperation of compliance which may then be assessed.

Promises have two polarities, with respect to this world: inwards or outwards from the agent, like a sender and a receiver of influence. These are the two basic viewpoints in a single interaction:

- Promises and impositions to give something (outwards from an agent) —denoted (+).

- Promises and impositions to receive something (inwards to an agent) —denoted (-).

In addition, third party observers may interpret something else entirely from the promises, giving a third view. If one agent promises to give something, this does not imply that the recipient agent promises to accept it, since that would violate the principle of autonomy. Similarly, one might impose on an agent to give something (please contribute to our charity), or to receive something (you really must accept our charity). Neither of these need influence the agent on which one imposes these suggestions, but one can try nevertheless.

Agents privately assess whether they consider that promises have been kept, in their world. They also assess the *value* associated with a promise, in their view. A promise to give something often has a positive value to the recipient and a negative value or cost to the agent keeping the promise, but not necessarily. An imposition often carries a cost to the recipient and brings value to the imposing agent. The existence of valuation further implies that there can be competition too. If different suppliers in a market make different promises, an assessment of their relative value can lead to a competition, or even a conflict of interest.

1.6 No special status for humans

Humans have a long tradition of trying to argue that we have been singled out as a special case in the world[4]; yet, each discovery in science reveals a simple truth: we are not as special as we often think, perhaps just a bit narcissistic! Many—if not all—the phenomena we associate with being especially human are indeed found elsewhere in the world, in some similar form, perhaps on a different level or scale. This is a powerful insight. It means we have a way to idealize and approximate all kinds of phenomena under a common umbrella of explanation, in both a qualitative and a quantitative way. That is the 'intent' of science.

Philosophers have, almost exclusively, tried to make intent and promises the exclusive domain of humans. In this book, we shall show that this is an unnecessary and even inappropriate response the concept—especially in this age of explorations into Artificial Intelligence, and the deeper understanding of all kinds of organisms in biology. Promises are related to intentions, and both are quite closely related to the idea of preferred outcomes. Consider the following examples:

Example 1. *A flight from London to New York promises onboard power. A British passenger took it for granted that she could plug in her phone or computer. When she reaches the flight, however, she finds that there are no British sockets and British voltages, only American plugs and voltages. Some seats have USB sockets. The promise was unclear and she failed to find a selection that brought about her intended outcome—to recharge her phone.*

Example 2. *Ten different contractors make slightly different promises about a service they could provide. Which one will you choose? Which one will you trust?*

Example 3. *Several chemical compounds can be used to treat a problem, but they have different costs, properties, and some will have side effects. How should we choose between them?*

Example 4. *During a flood, a river diverts to flow across the land, taking the path of least resistance and carving a new route to its goal: the ocean.*

The selections of 'goals' and 'outcomes' above are examples of what can be considered 'intended' behaviours. We tend to separate them because we unnecessarily and inappropriately attribute 'human agency' to these choices. In science, it's common to hear complaints about 'anthropomorphism', which is the use of language that attributes humanlike behaviours to inanimate objects, e.g. *'the particle wants to minimize its energy'*. Of course, it doesn't want anything—it doesn't have a brain. Nevertheless, such statements work and make complete sense to us because most such choices ultimately have nothing to do with conscious thought. They are simply possible choices—end of story. The attribution of humanlike qualities is harmless and irrelevant.

1.7 THE SIMPLE ESSENCE OF INTENT

If we sweep away the confounding sense of mystery—in bare minimal and practical terms—an *intention* is nothing more than the selection of a *possible outcome* from a number of alternatives, based on an optimization some criterion for success.

This statement essentially equates intent with a criterion, or a *type of information*. Forget about freewill, intent is something simpler and more scalable: a branching of choices into different possible worlds of outcome. Imagine being at a crossroads, and deciding a path. The path you choose will select your outcome, or your goal. Selecting a path is essentially the same as intending to reach a certain destination.

For example, we intended X but the unintended outcome was Y. This means we had a preference for X, but the causal uncertainties in the system led it to select Y. Labelling information is a key to doing this effectively.

Example 5. *Suppose you have a database that implicitly contains knowledge about the cure for infectious disease. If it isn't labelled in a way you can understand, you need a difficult process of trial and error to make that discrimination of data, and label everything symbolically.*

Once we've reduced all the noisy data to a simple symbolic discriminant, it's both cheap and reusable to promise which items are related to our particular branch of intent. So symbolic language is closely associated with intent and promises.

A simple lesson we can draw from these examples is that symbolism is a cost-saving compression of labelling information: like having an index of keywords to avoid having to search an entire book for an idea. Promises are labels that align with intended outcomes. This simple insight allows us to put promises to use immediately and see their effect on a wide range of scenarios, at all levels.

Example 6 (Taking for granted). *We 'take things for granted' when no promise of a behaviour has been made, but an agent makes use of the behaviour anyway, without considering if it will persist or not, or perhaps change ad hoc. Promises may be expensive to keep, so it is statistically likely that our expectations may exceed the promises offered, and that we might take stuff for granted in the face of 'incomplete specification'.*

The lesson of science it that we can make scenarios just as complicated and specialized as we want, but it's not helpful to drown in nuances. Idealizing and generalizing concepts, to a manageable level of simplicity, is nearly always more fruitful and pragmatic. In this book, we explore this style of reasoning to its logical conclusions.

1.8 THE HISTORY OF PROMISE CONCEPTS

In philosophy, the concept of promises has been discussed for the most part in the connection of morality. It comes with the implicit assumption that a promise implies an obligation (called the 'promissory obligation') to 'keep' the promise. There has been a recent revival of interest in this subject[Ati81, She11b, Sto52], at the time of writing[5].

Our basic departure from moral philosophy is that promises are not so much about morality as imposed onto the promiser, but about *information* and the management of expectation seen from the vantage points of both the promisee, and indeed any other individuals who might be privy to the information contained in the promise's content.

The concept of a promise is not a difficult one. It is a much simpler concept than obligations[BB13]. The tendency to favour obligation over promise originates in historical and theological convention, but this should not deter from studying promises as promises have both interesting theoretical properties and great practical value[6].

1.9 PROMISES IN INFORMATION TECHNOLOGY

A different motivation for promises was introduced by Burgess in the context of distributed management [Bur05]. Burgess uses the promise as a measure of information about 'voluntary cooperation' to circumvent fundamental problems with logics of obligation in determining system behaviour. Voluntary cooperation is a way of simplifying constraints and avoiding many-worlds paradoxes[Burb]. Any cooperation or even subordination of the parts that comes about in an organized system must then be understood as the result of 'honouring' purely voluntary promises to do so.

Voluntary cooperation can thus be used as a pragmatic engineering methodology for mapping out the behaviours of agents with both intentions and dynamical characteristics, in a way that is invariant with respect to centralization or decentralization of systems[7].

1.10 PROMISES AND TRUST ARE SYMBIOTIC

The usefulness of a promise is intimately connected with our *trust* in the agents making promises. Equivalently we can talk of the *belief* that promises are valid, in whatever meaning we choose to apply. In a world without trust, promises would be completely ineffective.

For some, this aspect of the world promise might be disconcerting. Particularly, those who hold to the notion of the so-called 'exact sciences'[8] are taught to describe the world in apparently objective terms, and the notion of something that involves human subjective appraisal feels intuitively wrong. However, the role of promises is to offer a framework for reducing the uncertainty about the outcome of certain events, not to offer guarantees or some insistence on determinism, and in many ways this is like modern theories of the natural world where indeterminism is built in at a fundamental level, without sacrificing the ability to make predictions.

1.11 How to use Promise Theory

To apply Promise Theory to a new problem, one starts in the following way:

1. *Identify the key agents.*

    Choose an agent to be any part of your system that can change or act independently. To get this part of the modelling right, one must be careful not to confuse intentions with actions or messages[9].

    To be independent, an agent only needs to think differently, i.e. have access to different information, or have a different perspective, etc. This may lead to distinctions between agents, which in turn may be used to enable functional specialization, and the 'separation of concerns' within a system. If we want agents that reason differently to work together, they need to promise to behave in a mutually beneficial way. In documenting this cooperation, one learns about the structure of intent in a system.

2. *Establish which agents make which promises.*

    The goal of cooperation is to ensure that agents make all the promises necessary so that some imaginary on-looker (the 'god's eye view'), with access to all the information, would be able to say that an entire cooperative operation could be seen as if it were a single entity making a single promise, i.e. one aims to establish whether the sum of the parts leads to a promise of collective behaviour.

    Instead of thinking in terms of 'requirements', we transform into a mirror image of autonomous promises.

    If agents are not already motivated to work together somehow, then we want to know how we coax the agents to make and keep promises that fit into the larger plan. How this works depends on what kinds of agents they are. If they are human, economic incentives are usually the answer. If the agents are programmable, then they need to be programmed to try to keep appropriate promises. Either way, we call this voluntary cooperation.

3. *Allow for the uncertainty.*

   Promises might or might not be kept, for a variety of reasons beyond the control of any agent, so plan for that. Even a machine can break down and fail to keep a promise, so we need to model this. Each promise will have a probability associated with it, based on our trust or belief in its future behaviour[10].

4. *Eliminate conflicts of intent.*

   Agents can only promise their own behaviour, so they can always resolve their own conflicts. The same is not true of obligations. On a larger scale, when agents work together, this requires more coordination. If all agents shared the same intentions, there would be no need for promises. Since the initial state of a system has unknown intentions, it means we have to set up things like 'agreements', where agents promise to behave in an aligned fashion. This is sometimes called orchestration of relative intent.

5. *Make assessments and valuations.*

   The outcomes of promises may be assessed by any agent in scope of the agents' activities. The role of the observer is key to understanding the essential *relativity* of agent interactions.

We hope to justify this summary in the coming chapters, with a semi-formal account. Our model has already proven very useful over these past two decades, and we hope that it will continue to be useful in the future.

Part I

Fundamentals

Chapter 2

The Heuristics of Promise Theory

Promises prime our expectations about the behaviours of things and people. They tell us what to expect before we have had the time to learn by experience. In other words, promises act as an anchor from which to start learning the behaviour of an agent, and establish its trustworthiness. In the inanimate world, promises express the 'invariant' and therefore expected aspects and quantities of other agents, as 'functional' characteristics (charge, mass, role, specialization, etc). Regardless of the sophistication and scale of the agents concerned, there is a role for the generalized notion of a promise and its 'intent'.

2.1 What qualifies as a promise?

Below are examples of the kinds of statements we shall refer to as promises. Let us begin with everyday statements and progress gradually to the kinds of abstract promises that we would like to use in a variety of technical scenarios.

- I promise you that I will walk the dog.
- I promise you that I fed your cat while you were away.
- We promise to accept cash payments.
- We promise to accept validated credit cards.
- I'll lock the door when I leave.

- I promise not to lock the door when I leave.
- We'll definitely wash our hands before touching the food.

These examples are quite uncontroversial. They are easily found in every day life, spoken by humans or posted on signs. It is easy to see that promises have certain characteristics. They are made voluntarily by an individual (the promiser or promisor[11]), to a recipient (the promisee), and they have a 'body' of description that details what the promise is about. We recognize that other parties may also know of the promise between promiser and promisees. These aspects will have to be captured in a model of promises.

2.2 PROMISES BY INANIMATE AGENTS

Although promises are of special interest to humans, humans do not need to be actively involved in order for promises to serve their purpose. Inanimate objects frequently serve as proxies for human intent. Thus it is useful to extend the notion of promises to allow inanimate objects and other entities to make promises. Consider the following promises that might be made in the world of Information Technology:

- The Internet Service Provider promises to deliver broadband Internet at a fixed for a fixed monthly payment.
- The security officer promises that the system will conform to security requirements.
- The support personnel promise to be available by pager 24 hours a day.
- Support staff promises to reply to queries within 24 hours.

These are straightforward promises, which could be made more specific. The final promise could also be restated in more abstract terms, transferring the promise to an abstract entity: "the help desk":

- The company help-desk promises to reply to service requests within 24 hours.
- The weather promises to be fine.

This latter example illustrates the way that we transfer the intentions of promises to 'entities' that we consider to be lifeless and incapable of forming their own ideas. They become responsible for keeping promises by proxy or association. It is a small step from this transference to a more general assignment of promises to individual components in a piece of technology. For example, we can document the properties of the following tools and technologies in the spirit of this argument:

2.3. A MODEL OF A PROMISE

- I am a meat knife and promise to cut more efficiently through meat.
- I am an electron, and I have charge $-e$ and mass m_e.
- I am a logic gate and promise to transform a TRUE signal into a FALSE signal and vice versa.
- I am a variable that promises to represent the value 17 of type integer.
- I am a command line interpreter and promise to accept input and execute commands from the user.
- I am a router and promise to accept packets from a list of authorized IP addresses.
- I am a compliance monitor and promise to verify and automatically repair the state of the system based on this description of system configuration and policy.
- I am a high availability server and I promise you service delivery with 99.9999% availability.

From these examples we see that the essence of promises is quite general. Indeed such promises are all around us in everyday life, both in mundane clothing as well as in technical disciplines. Statements about engineering specifications can also profitably be considered as promises, even though we might not ordinarily think of them in this way.

Example 7. *When an electronics engineer looks in a component catalogue and sees 'resistors' for sale promising to have resistance of 500 Ohms to within a tolerance of 5%, we do not argue about who made this promise or whether the resistor is capable of independent thought. The coloured bands on the component are a sufficient expression of this promise, and we accept it by association.*

By this reasoning, we propose that the concept of a promise should be formulated in way which allows for all of these uses. The value of this association is that promises are things that we use to form *expectations* of the behaviour of all manner of things. Such expectations contribute to reducing our *uncertainty* about their behaviour, and this can apply as much to technology as to humans.

2.3 A MODEL OF A PROMISE

Consider the following intuitive idea of a promise:

> *A promise is an announcement of fact or behaviour by a promiser about itself. The announcement is made to one more promisees, and may additionally be observed by some number of witnesses.*

The promiser, promisee and witnesses are assumed to be 'agents', and the promise is not assumed to be kept. It is merely a statement. This description fits the examples that we have already given and gives some clues as to the constitution of a promise.

The statement above opens up a number of questions that need answering. Already we can see that this apparently basic definition rests on a number of assumptions: that we can observe the outcomes of behaviours and that the outcome of a promise is clear at some single moment of time in the future, to be measured and verified by an observer. A full account of this might include a theory of measurement, but we might wish to avoid this level of detail as it binds us to too many details that have nothing directly to do with the issue.

The model world in which we formulate promises must have the following characteristics.

- There must be agents in order for promises to exist.
- There must be a promiser (or source agent).
- There must be a promisee (or recipient agent) which might be the same as the source.
- There must be a body which describes the nature of the promise.

We might summarize these attributes with a notation as in [Bur05]:

$$\text{promiser} \xrightarrow{body} \text{promisee(s)} \qquad (2.1)$$

- We can leave the body unspecified, but it must consist of a quality (a type, topic or subject for the promise) and a quantifier (which indicates how much of the realm of possibility for that subject is being promised). For example: promise quality: "travel to work", promise quantity "on Monday and Friday each week".

Finally, what is implicit in the above is that a promise requires the transmission of a message, or at least documentation in some kind of physical form, e.g. a speech act, or a written statement, else it cannot be made known to anyone except the promiser. A promise must therefore have documentation that is made intentionally or otherwise.

What then is a promise before we write it down? We shall refer to this as a *possible intention*. It might sound peculiar to use the this terminology for a passive attribute. For instance, if you say: "I promise that I am 20 years old", being 20 years old does not sound like an intention, but an attribute of the promiser. However, it is the subject of a possible intention to be something, even if you can't actually change what that is. 'Intention' is simply a word used in philosophy to capture all the possible things about which we can form an idea of something. If we promise to become 20 years old, it sounds less strange.

An intention is the basic formulation of a course of behaviour, which is made internally by (or on behalf of) an agent. When an intention is made public, it becomes a promise. If an intention is documented or leaked in some way then anyone has a right to assume it is a promise.

We take it as given that there has to be a source for every promise. A promiser does not have to reveal its identity of course, so witnesses to the promise might not know its source e.g. consider the anonymous threat. There is no reason to deny the existence of a source however. The lack of such information about a promiser is simply a defect in the knowledge of the receiving agent, but one would normally prefer to assume a consistent picture of promises and infer the existence of an anonymous promiser. This justifies our postulating the source. This implication of knowledge should indicate to us that relativity will play an important role in Promise Theory.

2.4 IMPOSITIONS

There is another class of interactions, in which an agent attempts to impose its intent on another agent to induce its cooperation without its prior consent. It includes suggestions, requests, demands, commands, and so on. We call such an interaction an *imposition*. Like a promise, it is not necessarily a successful attempt, but it is qualitatively different from a promise, and proves to be important in understanding the limitations and characteristics of a promise. We use the following notation for an imposition:

$$\text{imposer} \xrightarrow{body} \blacksquare \text{ imposee(s)} \qquad (2.2)$$

Impositions may try to induce behaviour that cannot even be promised by an agent, thus impositions are generally less likely to result in successful outcomes than promises simply because the receivers are unprepared for impositions. They may be unable or even unwilling to comply. Impositions may include commands, requests, referrals, laws, obligations of all kinds, and even *force majeur*. There is a long history of assuming that impositions, e.g. commands and orders, will be complied with in human and machine systems; that is a mistake. Promise Theory tells us that an agent's autonomy cannot trivially be breached without its direct promise of 'consent'.

2.5 IMPLICIT PROMISE VERSUS EXPLICIT PROMISE

The broadest and most far reaching usage of the term promise is in phrases like "the promise of solar energy" or "the promise of nano-technology". Such promises are implicit in the sense that the existence that a specific promiser is not assumed, and the word promise refers to an expectation or speculation of someone's assessment of what is being promised.

In contrast, an explicit promise from an individual comes by a promiser in an appropriate context. We will only focus on explicit promises below and we will assume by default that promises are explicit.

2.6 THE ROLE OF PROMISES TO REDUCE UNCERTAINTY

Promises lay the groundwork for *expectations* about the behaviours of things. Such expectations contribute to reducing our *uncertainty* about their behaviour, and this can apply as much to technology as to humans. We therefore take it as given that the concept of a promise is a useful one and consider next how one can formalize promises in the simplest and least assuming way.

Producing a promise may be more effective in reducing uncertainty than putting forward an assertion that is stated with more certainty. Indeed if an expectation about a piece of technology or about an agent is asserted with absolute certainty, or merely with some quantified probability of being valid, the question immediately arises how that knowledge has been obtained, thereby increasing uncertainty rather than reducing it.

Such existential questions do not arise for a piece of technology that has been delivered with promises to its users; they simply react to disappointing performance, perhaps losing trust in the promiser. Future promises from that same source would then be viewed with less credibility. Conversely, if a piece of equipment out-performs the promised performance, that fact may lead to increased trust in the original promiser.

2.7 OBLIGATIONS

Promises are often discussed in the context of agreements [She11a]. Invariably, however promises are linked with obligations. Although the promise concept has been mentioned in such diverse areas from logic, law and philosophy to economics, information science and computing, there is no agreement on what constitutes the semantic content of the terms or if there is even more than a tacit relationship between promise, commitment, obligation etc. For many, a promise implies a 'promissory obligation', and that is the starting point. More attention has been given to these concept of *obligations*, especially in the area of deontic logic. We believe, on the other hand, that the philosophical implications of promises are independent and far wider than is generally assumed, and that there is both a need and a practical importance to clarify them once and for all[12].

2.7. OBLIGATIONS

2.7.1 MEANING OF OBLIGATIONS

The intuitive notion of an obligation seems straightforward, but it proves to have difficult properties. We might try to think of obligations in a straightforward way, for instance: *an obligation is an intention that is perceived to be necessary by an agent.* This certainly captures some of the characteristics that we understand by the term, but it also leaves many questions unanswered: is the feeling of the necessity voluntary or forced, a matter of survival or simply an authoritarian convention? Another common understanding about obligations is that an agent should expect some kind of penalty for non-compliance with the obliged intention. Obligations thus seem to imply punitive actions.

2.7.2 SCOPE OF OBLIGATIONS

Obligations fall into the category of impositions, which are (of course) possible outcomes, and therefore possible intentions. Unlike promises, obligations have an ethical dimension and thus only occur in agents that partake in reasoning, e.g. humans. Agents that can't make decisions can not feel obliged to alter their behaviour (we can't shame a bottle into opening), but they may promise to get involved in certain kinds of interaction. This should not be limited to humans: biological processes and artificial machinery might also plausibly determine an obligation based on its promised behaviour. Obligations have source, a target, and a body, just like promises, but they are not implied by promises except under these very particular circumstances. The source and target are now somewhat difficult to understand however. We state some assumptions about them.

2.7.3 OBLIGATIONS AND FORCE AS POSSIBLE INTENTIONS

Obligations are imposed by an agent's exterior conditions, e.g. by the behaviour of external agents, perhaps by a threat of sanction, or simply by a force beyond its control, such as the weather. Obligations can also be self-imposed by codes of conduct, personal behavioural norms, which an agent holds to be *necessary*. This imposition suggests the action of a force which attempts to induce a commitment in another agent (or itself). An obligation is a possible intention, which may or may not be current, and may or may not have the status of a commitment. In any case, an agent is aware of the compelling reasons to include the intent to comply in its portfolio of commitments.

Even 'forced' behaviour can be classified under the realms of (possible) intentions, since all behaviour leads to outcomes and can thus be intended. Again, we emphasize that this does not imply that a coerced agent independently holds the intention that is being forced upon it. Nor does it say anything about whether the agent is able to resist the force or not, or whether it matters if an obligation is self-imposed or externally imposed.

Viewed from the perspective of an agent, the notion of an obligation immediately seems significantly more complicated than an intention or a promise, and does not seem to be close to the notion of either promises or intentions.

2.8 FOR AND AGAINST THE PRIMACY OF OBLIGATIONS

By sheer weight of tradition, obligations dominate discussions of behaviour. For that reason we make an attempt to compare promises and obligations as conceptual tools for distributed systems design.

2.8.1 IN FAVOUR OF OBLIGATIONS

1. Some people might think that a promise is an obligation because it seems to create one, and might therefore be considered equivalent to that obligation. (This is a version of obligationism which is opposed by later chapters.)

2. Obligations are a well known concept from deontic logic. There is an advantage to to reducing the less well-known concept of promises to one that has been studied for more than fifty years[13]. (This is true, but it implies no more than that promises are worth studying.)

3. Obligations have a formal status in state laws and regulations. There is no such public body of promises in the human realm, but there is so-called 'physical law' in the natural sciences, which is, in fact, has the status of a promise rather than an obligation.

4. Many obligations give rise to promises which occur in the process of fulfilling an obligation. E.g. the cat must get fed while owner is on holiday, the owner is obliged to get the cat fed (by law forbidding cruelty to animals). A friend promises to help in the fulfillment of the obligation. (This is true but it does not imply that obligations are prior to promises in general.)

2.8.2 AGAINST OBLIGATIONS

1. Suppose one has the concepts of promise available, and now reflects upon obligations. An agent $P_{issue:ob}$ issuing an obligation with body b_{ob} to all agents A in a scope S, might be understood as simultaneously promising all members A of S that $P_{issue:ob}$ will act in such a way as if P had received a promise with body B_{ob} from Q.

2.8. FOR AND AGAINST THE PRIMACY OF OBLIGATIONS 21

This is a reasonable explanation of what may happen when issuing an obligation and it explains why obligations seem to be complex entities or events from the perspective of promises.

2. If a future promise (e.g. the promise to feed the cat in the future) is in fact a deception then this falsifies the necessity of a relationship between promises and obligations. In other words, all intentions occurring as apparent intentions in promises cannot be induced obligations because some promises can be deceptions and these cannot be (easily) understood as induced obligations.

3. Similarly, not all promises are about future actions, so there cannot be an implied obligation capturing its essence for all promises. e.g. I promise that the cat got fed. Indeed the owner might actually be displeased that the cat was fed if it was supposed to be dieting.

4. In law, it is true that there is a dissimilarity between promises and obligations. They are quite different entities. Obligations may cause promises and promises may cause obligations, but promises have a physical reality as events in space and time, whereas obligations do not. Obligations are at a different level of abstraction altogether.

5. Promises are made on a voluntary basis. For obligation however, the concept of voluntary is almost irrational. In any case it might be voluntary to imply an obligation on someone else, but engaging in a promise you may face an involuntary obligation or a voluntary one. Voluntary-ness is therefore natural for promises but is quite problematic for obligations.

6. Promise announcement constitutes positive extensions of user behaviour, whereas obligations constitute a negative constraint on the degrees of freedom of the obliged party.

It seems clear that promises are a simpler concept than obligations. The concept of a promise is more natural in the natural and technological world. Computers cannot feel ethical responsibility, hence the reduction to promises to obligations seems to be neither philosophically satisfactory nor technically correct. From a pragmatic perspective, there are some behaviours one cannot oblige (empty the ocean with a sieve); however, these can be promised, even if the promises are clearly invalid. Thus, we maintain:

- Promises are different, simpler and can be analyzed independently of obligations.

- Promises are *local* constructions, whereas obligations are *non-local*.

The source of a promise is localized in a single entity that has all of the information and self-control to be available to resolve conflicts and problems with multiple promises. The sources of obligations however are distributed amongst many individuals and the obliged party does not have the access to resolve the conflicts without maintaining a voluntary dialogue with all of these multiple parties.

- Obligations tend to increase uncertainty not reduce it. Obligations can be inconsistent, but promises cannot. More precisely: consistency of promises is a matter that can be verified at the level of sources only. Promises made by different agents cannot be inconsistent.

One would prefer not use obligation as a coordination principle, if a mechanism based on promises could be used instead. Promises are simply more trustworthy. A collaboration based on promises works better if one has trust. In a world of obligations however, trust is meaningless because one has only a presumed outcome.

2.9 Assessing promises

The notion of whether promises are kept or not is central to their sustained usefulness, thus we need to make mention of how this comes about in a theory of promises. It would be easy to go overboard and delve into the complexities of observation and measurement to provide a satisfactory answer but that is not in the spirit of this work. We seek instead a simpler notion which is at the same level of abstraction as the concepts of promise and intention that we have introduced thus far. We call this the concept of *assessment*.

> *An assessment is a subjective statement made by an agent about whether the intentions of itself or of another agent were fulfilled.*

The notion of an assessment is generic, not necessarily quantitative. It is both subjective and *a priori* unlinked to an actual observation. At this level of description, we need not say any more about it than this. The 'reality', according to some agent, of whether promises have been kept or not is neither here nor there. What is important is how some agent assesses the state of the promise for its own purposes.

Some examples of assessments are shown in table 2.1. We see that assessments are quite sensitive to physical representation of the promise. Once again the notion of representation (or documentation) is a key to the importance of a promise as a concept.

Promise	Representation	Scope	Assessment
Fed the cat	Speech act	Those who heard	Either did or did not.
Credit card accepted	Action	Visitors to the store	Either did or did not.
Response in 24 hours	Contract	Signing parties	Replied in time or not.
$var = value$	Source code	Readers of the code	Syntax ok. Value in range.
$var = value$	Object code	Execution engine	Ok to execute, exception, etc.

Table 2.1: Some example promise assessments

2.10 THE IMPLICATION OF ASSESSMENT

The fact that promise outcomes are observable events has deep implications for the interpretation of promise modelling. The act of assessment, through measurement and by an observer, implies a relativistic model of time. The act of measurement is a sampling of information, which is a tick of the observer's clock. It implies the existence of an interior *process* for each agent, with independent dynamics, that is capable of noticing a change in the agent whose outcome is being assessed. As such, Promise Theory is actually a theory of implicit *processes*. No static agent can detect changes or therefore outcomes over time (for a popular discussion of this issue, see [Bur19b])

The issue of observability has often been confronted in physics, e.g. in relativity and quantum mechanics, where it's sometimes claimed that observer processes have to be conscious and even intelligent. This is false. However, Information Theory does tell us that any agent sampling the activities of another must have an independent time of its own, which ticks away at the Nyquist-Shannon sampling rate for the maximum change rate it can discriminate. Promise Theory thus has a similar reliance on observability as any other interaction theory. However, rather than assuming that the observer is a human, Promise Theory treats any agent as an observer, on any scale. This brings a simple clarity to the meaning of observation, and avoids the need to speculate about the consciousness and freewill of observers.

2.11 THE VALUE OF PROMISES

Promises are valuable to agents for many reasons.

- They help reduce uncertainty.

- Their outcomes could be beneficial.

- Actual transactions (e.g. money) might be involved.

Even the presence of a promise, before it has been kept, can be valuable as it allows agents to plan ahead. Because certainty is key, a promise adds nothing unless there is trust. Zero trust makes promises worthless. Trust usually begins at some default level and is increases or decreases based on a history of keeping promises—or, in our terminology, on a history of positive assessments about a succession of promises. The value of a promise is an assessment made by each agent individually. It could be based on the expectation of the eventual benefit.

Example 8. *Suppose agent A promises agent B 400 dollars per year. B promises to wash A's windows at this price. Both are satisfied with the value they get from this arrangement and prefer not to question it too much as this could unleash all kinds of consequences. Observer C can see that the values are quite mismatched, or that A is getting a poor deal by its judgement, but C also cannot deny that the relationship is stable because both A and B are happy.*

2.12 DECEPTIONS AS PROMISES

Simple agent behaviour is complicated mainly by probabilistic outcomes, but more complex agents can engage in complex reasoning and motivation. In order to scale to all kinds of scenario, we need to allow for this. Understanding deceptions (or lies) is also an important step in clarifying the relationship between intentions and promises, because it is possible for an agent to have two different intentions in play at the same time: an actual intention to which it is committed and an announced intention (i.e. a promise) which are not compatible. Incompatibility means that striving for both intentions simultaneously is fruitless because their realizations cannot be combined. In that case, we can speak of a deception.

In a deception, the hidden intention is more important than the witnessed one one and we might refer to it as the dominant intention. This simply expresses that it is a commitment while the promise contains merely a "possible intention". It is the *real* intention of the agent ("intended intention"), while the intention in the promise can merely be described as *non-realized*. If the dominant intention should be rescinded, a deception will revert to being a promise, but this is only known to the source.

We begin to appreciate why the concept of a promise is in fact so important. A promise is simply a documentation of an intent, regardless of what lies behind it. Any internal priorities or considerations are hidden from the view of other agents and cannot be observed. Thus, promises are an independently important concept because we can (indeed must) talk about promises without discussing the basis on which they are made.

When a promise is made, we are neither required nor able to confront the truth or falsity of the promise. Indeed, as soon as we ask such questions, new issues such as trust and a plethora of other subjective issues come into play. Such issues are probably un-resolvable in a logical sense. However, what we assume is here that no matter how trustworthy a promise might be, it can increase or decrease our certainty of a promised outcome and thus it bears an *influence*. The matter of assessing the promise can be very complicated and uncertain and we shall not attempt to discuss this here in any depth.

2.13 More elements of Promise Theory

Throughout this work, we shall be confronted with issues from two different perspectives: dynamical and semantic[14].

- The *dynamics* of agents and their promises refer to actual measurable patterns of behaviour, describable in terms of space and time and the agents' characteristic variables.

- The *semantics* of agents refer to agents' interpretations of one another's behaviour, whether intentional or not. The semantics of a promise or unpromised behaviour is always an assessment made by an observer. "The observer is always right" in an autonomous view.

In addition, there are topics we simply cannot cover in this introduction, such as the economics of what motivates agent interactions, and the links to many other related fields of description in mathematics, physics, and others. These fall outside the scope of the present work, but some have been addressed in other interaction theories including but not limited to Game Theory[Mye91, Ras01] and Category Theory[Pie91].

CHAPTER 3

FORMALIZING PROMISES

We now turn to a specific model to capture the essential features of promises. We originally called this *micropromises*, or μ-promises, to emphasize that it is a particular formalization of a promise[15]; however, this no longer seems necessary as the model has not been challenged, so we shall henceforth refer to this model exclusively when we speak of promises, unless otherwise stated.

3.1 AGENTS

The only entities in a world of promises (or μ-promises) are *autonomous agents*. Agents play the role of *promisers* when making promises; in the role of *promisee* they are the recipients of such promises. In the role of *observers*, agents merely receive messages and signals from other agents and *assess* whether promises are kept. Agents are not generally observable by one another unless they promise to be.

The internal workings of agents are assumed to be hidden from one another, unless knowledge of their details is promised. Agents may be composite entities, composed of several lesser agents, which together make promises. Observers may or may not be able to know this. Table 3.1 shows some examples of real world entities that could be considered promise agents.

3.2 FORMAL (MICRO)PROMISES DEFINED

We denote agents by upper-case Roman letters. A set of N agents is written A_i, individually where $i = 1, \ldots, N$, and collectively $\{A_i\}$. We shall often use nicknames A_s for

3.2. FORMAL (MICRO)PROMISES DEFINED

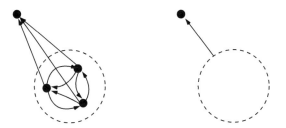

Figure 3.1: Agents might have internal structure (left), revealed by differentiated promises, but that structure might be presented as a single agent package (right). The atomicity of the agents is therefore not as fundamental as the atomicity of the promises they make.

Agent	Promise body
Electron	Negative electric charge
Web server	Deliver HTML pages
Can of soup	Contains soup
Bus	Transport to location on front sign
Human	Obey the law of the land.

Table 3.1: Some example promiser agents.

promise-sender or giver, A_r for promise receiver and A_t for third parties, to assist the discussion.

Definition 1 (Promise or μ-Promise). *A promise is an autonomous declaration of intended, but as yet unverified, behaviour from one agent (the* promiser*) to one or more others (called* promisees*). Each promise contains a body b that explains what is being promised.*

We denote a promise from agent A_s to an agent A_r, with body b by:

$$A_s \xrightarrow{b} A_r \tag{3.1}$$

A promise refers to unverified behaviour, because one does not promise a state of affairs that is already known or does not need to be verified again. It can refer to events in the past or the future, but information about the state that is being promised has yet to be verified.

Example 9. *One might promise to pick up someone from the airport, about an event in the future, or promise to have finished an assignment in the past. The key point is that the recipient of the promise has yet to verify the outcome.*

3.3 Promise proposals and signing

Promises can exist and be discussed without them ever having been made or intended to be kept by an agent. We have used the notation $\mathbf{def}(\pi)$ for a description of the promise π. This description must be adopted by an agent before it can be said to have made the promise. It is therefore useful to maintain the idea of promise proposals.

> **Definition 2** (Promise proposals). *The statement of a promise that is posited for consideration by one or more parties, prior to keeping or discarding the promise.*

In other words, a proposal is a complete description of a possible promise, with the caveat that this is not yet intended. Promise proposals are often discussed as part of treaty negotiations and commercial relationships, such as contractual relationships.

3.3.1 What we can say about the body?

The content of the promise is referred to as the body of the promise, by analogy with the body of a document, or the body of a contract. The body explains the extent to which the making of the promise either *will constrain* or *has constrained* the behaviour of the agent making it.

A body b may contain:

1. A name or label $\Lambda(b)$ that uniquely identifies the promise from similar of dissimilar promises to the agent.

2. A type $\tau(b)$ that explains the nature of the promise to the agent.

3. An explicit constraint $\chi(b)$ on the affected state of the agent.

It is helpful for promises to have names so that they may be distinguishable from one another. This enables both provenance and distinct causation. Although the promised outcome can never depend on the names of promises, it is often helpful to have a unique name when promise bodies might overlap[16].

The constraint $\chi(b)$ might not obviously be a constraint.

Example 10. *If you promise to brush your teeth, you are really promising one course of action out of all the many possible outcomes of your behaviour of type 'state of teeth'.*

A natural way is to classify promise types by the kinds of entities that are affected. This is up to each agent, as an autonomous entity to decide. We might then define the promise type $\tau =$ 'teeth' with set of possible outcomes $o \in \{white, clean\}$. Then the promise to brush your teeth is involves the the desired outcome constraint $\chi(o) : o = white$.

3.4. NOTATION AND CONVENTIONS

Promise types distinguish the qualitative differences between promises, and additional constraint attributes distinguish the quantitative differences. Thus we shall often write the promise body as a pair $b \sim (\tau(b), \chi(b))$.

1. For each promise body b, there is another promise body $\neg b$ which represents the strict negation of b. We assume that $\neg\neg b = b$.

2. We further assume that $\tau(\neg b) = \tau(b)$, i.e. the type of promise is not affected by its constraint, and $\neg\chi(b) = \chi(\neg b)$, i.e. negative applies only to the constraint.

3. The negation of b refers to the deliberate act of not performing b, or what ever is the complementary action of type $\tau(b)$.

3.4 NOTATION AND CONVENTIONS

We use the term *promiser* for the giver of a promise, as opposed to the legal term *promisor* often used in the moral literature of promises. When we indeed want to express an equivalence, we use the terms:

$=$ is equal to, becomes
$\equiv$ identity to
$\sim$ approximately the same as

- We use the vertical bar | as in conditional probability to mean 'if' or 'given that', so $b|c$, means 'b given that c is true'.

- The symbol $\vee$ represents 'OR' (think inclusiveness going up).

- The symbol $\wedge$ represents 'AND' (think inclusiveness going down).

- We denote a promise body b subject to a conditional expression c by $b|c$.

- We denote the truth of a conditional expression c by $T(c)$.

- We denote the falsity of a conditional expression c by $F(c)$.

- $\gtrsim$ may be read as 'reduces to', or 'has the effect of' or 'is functionally equivalent to'.

Particular agents may be written by name rather than using a symbol A.

- A_i is the i-th agent of a number of agents identified by their label. These may or may not be distinguishable from one another.

- A promise is referred to collectively with the symbol π, or

$$\pi : A \xrightarrow[\sigma]{b} A', \qquad (3.2)$$

for promiser A, promisee A', body b, and scope σ.

- We denote knowledge of a promise, or its definition by $\mathbf{def}(\pi)$, which is essentially a tuple $\langle A, b, A'; \sigma \rangle$.

- A promise directed to 'any agent' is written to $A_?$, i.e. $A \xrightarrow{b} A_?$.

- A promise directed to 'any agent' or all agents may be written to $*$, i.e. $A \xrightarrow{b} *$.

3.5 SCOPE, WORLDS, AND RELATIVITY

The fact that each agent is an autonomous entity means that it observes and stores information in a unique perspective, which is *a priori* isolated from other agents. Each agent's view of the world of other agents is thus initially unique.

If two agents attempt to compare their information or world views on a particular subject, they might or might not agree. This is called agent relativity[17]. The relativity of agents will force us to confront so-called *many-worlds* scenarios[18], in which every agent experiences, for all intents and purposes, a completely independent branch of reality, until is begins to cooperate and exchange information with other agents.

3.5.1 DENOTING INFORMATION ABOUT PROMISES

Agents need to refer to the promises they make. Knowledge of a promise could itself be the subject of a promise (e.g. I promise that X told me about her promise to be faithful to Y), so there must be promise bodies that contain the knowledge of other promises. We shall need a standard notation to represent this[Bur09].

The notation π_i ($i = 0, 1, 2, 3, \ldots$) labels complete promises including the promiser, a list of promisees, and the body with all of its details. We shall write the passive definition of a promise as by $\mathbf{def}(\pi)$. This is meant to denote a distinction between a passive depiction of the promise $\mathbf{def}(\pi)$ and the active realization π, just as there is a different between a newspaper article about a promise and the promise itself.

Definition 3 (Knowable information about promises). *The promise body for talking about a promise π is denoted by by $\mathbf{def}(\pi)$. We define $\mathbf{def}(\mathbf{def}(\pi)) = \mathbf{def}(\pi)$.*

If we wish use this as part of a promise, e.g. I promise (π_1) that I promised something else (π_2), then the first promise π_1 has body 'promise description' $b_1 = \mathbf{def}(\pi_2)$. In

3.5. SCOPE, WORLDS, AND RELATIVITY

other words, promise π_1 constrains the body of the promise to simply be a verbatim description of the promise π_2, like packaging in an envelope.

3.5.2 DENOTING PROMISE SCOPE

Knowledge about promises is at the heart of the notion of promise *scope* (see section 2.3). The ability for a promise to drive expectations relies on it being known. The scope of a promise is the collection of agents that have received information about its existence.

Definition 4 (Scope). *We denote the scope of a promise by a set of agents σ, with whom information $\mathbf{def}(\pi)$ is shared. This scope may optionally be written under the promise arrow.*

$$A_1 \xrightarrow[\{\sigma\}]{b} A_2 \tag{3.3}$$

The default scope is the promiser plus the promisee:

$$A_1 \xrightarrow{b} A_2 \equiv A_1 \xrightarrow[\{A_1, A_2\}]{b} A_2 \tag{3.4}$$

3.5.3 TYPES AND LABELS

We clearly benefit from differentiating *types* of agents, but where do these types come from? A priori, agents have no type: they are homogeneous, structureless, universal elements (analogous to biological 'stem cells') that may only be differentiated via the promises they make. A *type* may thus be defined either by identification of a role (see section 4.4), or by explicitly making a promise.

Consider a universal (typeless) agent $A_\emptyset$, that makes no initial promises, or empty promises to everyone:

$$A_\emptyset \xrightarrow{\emptyset} *. \tag{3.5}$$

We may add a promise, with additional scope σ:

$$A_\emptyset \xrightarrow[\sigma]{+\text{fish}} * \tag{3.6}$$

Any agent in the scope $\{*, \sigma\}$ may now identify the former agent as being equivalent to an agent of type b, making no promise. In other words, within the scope, the agent effectively has a new name:

$$A_{\text{fish}} \xrightarrow[\sigma]{+\emptyset} *, \tag{3.7}$$

thence

$$\text{fish} \xrightarrow[\sigma]{+\emptyset} *. \tag{3.8}$$

i.e., we can drop the promiser's agent designation 'A', and label agents simply by their promised names. We also note that promises that do not explicitly make unique promises are therefore indistinguishable. The name of a 'type' therefore describes a role by association (see section 4.4). We use this notion from here on to write agent types implicitly, e.g.

$$T_1 \xrightarrow{\emptyset} A \equiv A_1 \xrightarrow{+T} A \tag{3.9}$$

$$R_2 \xrightarrow{\emptyset} A \equiv A_2 \xrightarrow{+R} A \tag{3.10}$$

$$H_3 \xrightarrow{\emptyset} A \equiv A_3 \xrightarrow{+H} A \tag{3.11}$$

and so on. In this way, we can move labels from the promise body to the agent's identifier at will. This is in keeping with the idea that the agent's name is the basic promise that identifies it.

3.6 IMPOSITIONS

Promises play an important role in autonomous cooperation, but not all intentional behaviour occurs spontaneously from within agents. We also need a way to model things like suggestions and requests, where one agent attempts to impose on another's autonomy to indicate a behaviour it desires. We call such communications *impositions*, as they challenge the autonomy of other agents in some respect.

Definition 5 (Imposition). *A message intended to induce voluntary cooperation in another agent. We denote an imposition from agent A_1 to agent A_2 by the symbol (imagine a fist):*

$$A_1 \xrightarrow{b} \blacksquare\, A_2 \tag{3.12}$$

Impositions have scope in the same way as promises.

Impositions come in varying degrees of strength, for example:

- Hints and suggestions.
- Requests and proposals
- Requirements and specifications.

- Commands and demands.

Impositions are not promises because they cannot be kept by the agent who makes them. Moreover, impositions are not quite obligations, as the latter imply some form of penalty for non-compliance. We may think of them as a desire to impart intentional behaviour in an external entity.

3.7 OBLIGATIONS

Obligations are 'deontic' statements like 'X must comply with Y', 'A should do B', 'C is allowed to do D', and so on. Obligations are imposed on an agent from outside, i.e. by external agents. Autonomous agents are, by definition, never obliged to do anything they have not decided for themselves. To accept an obligation, an autonomous agent must sacrifice some of its autonomy (see the sections beginning with 6.3). This assumes the notion of *authority* (see section 13.8).

There are generally three agents involved in an obligation: an agent imposing the obligation (an authority), and agent that is obliged to promise something to another and an agent and the receiving agent.

3.8 KINDS OF PROMISES

Autonomy implies that agents should never make promises on behalf of other agents, as the other agents would always be free to ignore such impositions. This de facto rule leads to a highly desirable simplicity in μ-promises: it leaves agents a priori decoupled from one another. However, it is possible to formulate more general promise-like assertions, and we wish to show that these additional levels of complexity do not add real expressive power to a theory in which one has only promises and impositions.

Below we describe four kinds of promises, in increasing stages of complexity. The generalized notation for these promise may be written as an extension of the earlier notation:

$$A[B] \xrightarrow{b} C[D] \tag{3.13}$$

where A is the promiser, C is a promisee or intended recipient of the communication, B is the agent assigned responsibility for keeping the promise, and D is the party affected by the outcome.

3.8.1 PROMISES OF THE FIRST KIND

The simplest form of a promise is the one we have used up to this point, and which we call μ-promises. We write it:

$$\pi : S \xrightarrow{b} R \qquad (3.14)$$

where π is its name, and the triplet $\langle S, b, R \rangle$ represents the sender agent S, the promise body b, and the recipient agent R.

It is assumed that this is a local and autonomously made promise. It is equivalent to a promise of the fourth kind, described by the notation:

$$\pi : S[S] \xrightarrow{b} R[R]. \qquad (3.15)$$

i.e. S promises b to R.

3.8.2 PROMISES OF THE SECOND KIND

A promise of the second kind allows an imposition:

$$\pi : S[T] \xrightarrow{b} R \qquad (3.16)$$

i.e. S promises R that it will impose on T to act as if it had promised b to R. If T is autonomous, this is forbidden and has no influence on T. Using impositions and promises of the first kind, we may write this:

$$S \xrightarrow{T \xrightarrow{b} R} \blacksquare T. \qquad (3.17)$$

Note that, in this description, we do not distinguish between obliging and imposing. This is a subtlety yet to be resolved. Ultimately, the notion of autonomous intention would have to be sacrificed in the presence of overwhelming force.

3.8.3 PROMISES OF THE THIRD KIND

A promise of the third kind is autonomously given but allows indirection.

$$\pi : S \xrightarrow{b} R[T] \qquad (3.18)$$

i.e. S promises to R that S will do b for T. Promises of the third kind hold a special significance for transmission of knowledge and promise *scope* (see section 3.5).

Consider a promise by S to R about a promise with body b to a third agent T. We could write this in two ways. We use the notation $\mathbf{def}(b)$ for a promise body that

3.8. KINDS OF PROMISES

describes the knowledge about promises bodies of type b. So we could write this as a promise of the third kind directly:

$$\pi_3 : S \xrightarrow{b} R[T] \qquad (3.19)$$

which would furnish R with complete information about b and its recipient T. A similar effect could be attained by writing this as a promise of the first kind about the knowledge $\mathbf{def}(\cdot)$ of the original promise:

$$\pi_1 \; : \; S \xrightarrow{b} [T] \qquad (3.20)$$

$$\pi_2 \; : \; S \xrightarrow{\mathbf{def}\,\pi_1} R \qquad (3.21)$$

This is not quite the same thing, because now T is also in the scope of the promise, being the promisee, while it was not in the original version. In fact, all (3.19) really expresses is the fact the benefactor of the promise T is communicated as part of the body of the promise, so we could equally write:

$$S \xrightarrow{b(T)} R \qquad (3.22)$$

This is a promise about a promise. Clearly, we cannot exclude the names of agents from the promise body in general.

3.8.4 PROMISES OF THE FOURTH AND MOST GENERAL KIND

A promise of the fourth kind is the most generic form:

$$\pi : S[T] \xrightarrow{b} D[U] \qquad (3.23)$$

i.e. S promises D that T will act as if it had promised b to U. Note that, in this case S, T cannot be considered an autonomous agent, for if T had been autonomous, this would be forbidden as S would have no influence of T. However, once again, we can express the intention here using purely voluntary cooperation, using promises of the first kind – this time, it involves a promise about a promise.

There are two possible interpretations of this. Keeping the notational meaning from (3.17), we may write this as a promise by S to D that S imposes on T to make a promise to U with body b.

$$S \xrightarrow{S \xrightarrow{T \xrightarrow{b} U} \blacksquare T} D \qquad (3.24)$$

Another less satisfactory interpretation, which assumes that the imposition is accepted and promised by T, can be written purely in terms of promises of the first kind is the following:

$$\pi_1 : S \xrightarrow{\pi_3} D \tag{3.25}$$

$$\pi_2 : S \xrightarrow{\pi_4} D \tag{3.26}$$

$$\pi_3 : T \xrightarrow{-\mathbf{def}(b)} S, \tag{3.27}$$

$$\pi_4 : T \xrightarrow{b} U, \tag{3.28}$$

i.e. S promises D that it knows of a promise π_3 (which has body $-\mathbf{def}(b)$) made from T to S to acquire information about body b (see section 3.11.1). S knows this information according to equation. 3.23. Moreover, S knows of another promise π_4 made by T to U that it intends to use this knowledge of b to make a promise itself with this body to U.

3.8.5 Elementarity of promises of the first kind

From the preceding cases, it appears as though we can construct facsimiles of all of the generalized promise kinds, that flout the strict rules of autonomy, using only promises of the first kind. This means that we can always re-express in terms of promises of the first kind and impositions. This motivates the study of these elementary promises along with impositions with no loss of generality.

Theorem 1 (Projection theorem for promises of the n-th kind). *Let $\pi^{(n)}$ be a promise of the n-th kind, and A_1, A_2, A_3, A_4 be agents that are not necessarily, distinct. It is always possible to substitute for $\pi^{(n)}$ a finite number of promises of the first kind, along with impositions, such that the resulting assurances are the same.*

The proof is contained in the discussion of the cases above. Henceforth, the promises in this book shall only refer to those of the first kind, and impositions will be treated as a separate phenomenon.

3.9 Promise graphs and networks

The theory of graphs and networks is an important part of mathematics and indeed science, and was influential in formulating Promise Theory. Visualizing promises in terms of graphs or networks helps us to see even implicit channels of influence explicitly. However, it requires some care, as the channels of influence follow the scope of promises rather than just the obvious promiser to promisee route.

Our formulation of μ-promises has the obvious characteristics of a network, or in graph theoretical terms a so-called *directed graph* (a network of arrows). This is not a novel or unusual construction[19]; many phenomena form such networks. However, its commonality is a powerful identification of its ubiquity, and this feature of promises will allow us to draw in many important insights that have been made about networks in later chapters. For example, directedness in graphs can display *causation*.

When a single agents makes a collection of promises to other agents, some of these can often be simplified or replaced by a single promise made to a collective 'super agent'. There can also be cases in which we attribute special meaning (semantics) to particular combinations of promises, thus we begin by discussing the basics of pattern composition.

The description of graphs is closely related to the concept of matrices that describe graph connections. The adjacency matrix for a graph is a table of rows and columns that are labelled with the agents or node locations in the graph. The table contains numbers, like a spreadsheet, which represent the existence of a connection between the agents in the row and the column. If there is a link between one agent and the next, we say that they are *adjacent*. The table is called the adjacency matrix.

3.10 DEFINING PROMISE MATRICES

It is useful to define two matrices:

Definition 6 (Promise matrix). *Let i, j label a collection of n agents, where $i = 1, \ldots, n$. The promise matrix π_{ij} is the matrix of all promises between A_i, A_j, stripped of the agent labels, i.e. in which the agents are implicit. Hence, the complete set of promises between the agents may be written:*

$$\bigcup_{i,j=1}^{n} A_i \, \pi_{ij} \, A_j = \sum_{i,j=1}^{n} A_i \, \pi_{ij} \, A_j, \qquad (3.29)$$

This notation defines the meaning of $\sum$ and $+$.

Definition 7 (Promise adjacency matrix). *The directed graph adjacency matrix which records a link if there is a promise of any type between the labelled agents.*

$$\Pi_{ij} = \left. \begin{array}{l} 1 \quad \text{iff } A_i \xrightarrow{b_{ij}} A_j, \\ 0 \end{array} \right\} \quad \forall b_{ij} \neq \emptyset \qquad (3.30)$$

These may be further decomposed into useful subsets, for instance, the rank decomposition of the promise matrix, into matrices of promises of rank r, will be useful:

$$\Pi_{ij} = \sum_{r=0} \Pi_{ij}^{(r)}. \qquad (3.31)$$

The concept of an agent's interior and exterior promises will also be defined below.

3.11 Transfer of intentional behaviour between agents

In order for intentions and behaviours to cross the autonomy barrier from the domain of one agent to the autonomous domain of another, agents on the receiving end of promises and impositions have to (autonomously) intend or promise to accept this.

On a directed graph, promises are represented as incoming or outgoing arrows, pointing from promiser to promisee. It follows from the autonomy of agents in a system, that for every promise to *give* or *provide* something (represented by a + sign) from A_1 to A_2, there is a possible counter-promise to *accept* or *receive* (represented with a − sign). Such pairs of promises form the basis of *negotiations*, and a pair of promises back-to-back are said to form a *binding* (see section 8.4).

> While a promise is a directive to oneself, an imposition is a directive to another; thus the polarities of promises and impositions made by an agent are always relative to recipients at opposite ends of a cooperative relationship.

3.11.1 Polarity ± of promises

It is helpful to denote the flow of intent using signed promises. We shall denote promises with a ± signs, as below:

$$A_1 \xrightarrow{+b} A_2 \quad \text{(I will give } b\text{)} \tag{3.32}$$

$$A_1 \xrightarrow{-b} A_2 \quad \text{(I will accept } b\text{)} \tag{3.33}$$

The significance of these two classes of intent is fundamental and reveals a basic duality of viewpoints when formulating collective behaviour in terms of promises (see section 3.11.6), not unlike the way that positive and negative charges lead to crucial insights about electricity.

3.11.2 Promises to use

Merely labelling a promise body $-b$ is ambiguous, so we shall take it as given that nothing more than acquiring of information is implied by such a promise. A promise with body $-b$ is a specification of what behaviour will be "received" or not blocked by the agent.

We shall sometimes emphasize the promise to use another agent's promise of a service by using a more visible notation:

3.11. TRANSFER OF INTENTIONAL BEHAVIOUR BETWEEN AGENTS

$$A_1 \xrightarrow{S} A_2 \text{ or } A_1 \xrightarrow{+S} A_2$$
$$A_2 \xrightarrow{U(S)} A_1 \text{ or } A_2 \xrightarrow{-S} A_1 \tag{3.34}$$

The body notation $U(S)$ will sometimes be used to denote a use-promise; often we simple use the $-$ sign. The effect of using another use-promise is to actually provide the behaviour the other is trying to use. Thus we can note the functional equivalence of:

$$\left.\begin{array}{c} A_1 \xrightarrow{U(b)} A_2 \\ A_1 \xleftarrow{U(U(b))} A_2 \end{array}\right\} = \left.\begin{array}{c} A_1 \xrightarrow{-b} A_2 \\ A_1 \xleftarrow{--b} A_2 \end{array}\right\} \Rrightarrow \left.\begin{array}{c} A_1 \xrightarrow{-b} A_2 \\ A_1 \xleftarrow{+b} A_2 \end{array}\right\} \tag{3.35}$$

This is not reversible, however. Moreover, one must be careful with the shorthand pseudo-functional notation $U(b)$. It is tempting to want to write that $U(U(b)) = b$, however this incorrect. The body alone does not satisfy this property in any meaningful way. The equivalence of behaviour is between the promises, which involves multiple agents and promises in both directions. This cannot be represented by a condition on the body alone.

3.11.3 THE USE PROMISE AS A NON-PRIMITIVE

The use-promise we have referred to so far cannot be primitive promise type, since it includes implicit information about the promise. We can express this by defining:

$$\left.\begin{array}{c} U(b) \\ -b \end{array}\right\} \equiv \text{acquire } \mathbf{def}(b) \text{ and } \mathbf{employ}(b). \tag{3.36}$$

i.e.

$$\text{Use} \equiv \text{knowledge of content } b \text{ , intention to employ the content } b$$
$$\text{somehow.}$$

The notation $U(b)$, like $-b$ is not without some ambiguity. It indicates the directionality but we still need to clarify its intent. We recommend using it as a shorthand for a proper description of its meaning.

3.11.4 POLARITY OF $\pm$ IMPOSITIONS

By analogy with (3.33), impositions can also refer to giving and receiving of behaviour.

$$A_1 \xrightarrow{-b} \blacksquare A_2 \text{ (You give } b!\text{)} \tag{3.37}$$
$$A_1 \xrightarrow{+b} \blacksquare A_2 \text{ (You take } b!\text{)} \tag{3.38}$$

Although there are two polarities to impositions, providing a certain symmetry with promises, there can be no corresponding 'use' interpretation for an imposition, because an imposition always refers to a directive for another agent, whose behaviour is autonomously determined.

> Promising to receive something, and imposing on something or someone to receive something are not equivalents. The promise is made by the agent in charge of its autonomy, while the imposition is not.

3.11.5 COOPERATIVE INTERACTIONS BETWEEN AGENTS

Autonomous agents only interact fully if two or more parties intend to do so. There are several ways in which intentional behaviour can influence pairs of agents[20].

1. A_1: I promise you b; A_2: I promise to accept b.

$$\left. \begin{array}{c} A_1 \xrightarrow{+b} A_2 \\ A_1 \xleftarrow{-b} A_2 \end{array} \right\} \quad (3.39)$$

2. A_1: Take b! (imposition) A_2: Okay, I promise I will.

$$\left. \begin{array}{c} A_1 \xrightarrow{+b} \blacksquare A_2 \\ A_1 \xleftarrow{-b} A_2 \end{array} \right\} \quad (3.40)$$

The acceptance of an imposition may appear as a mandate to justify it (see section 13.8).

3. A_1: Give me b! A_2 (demand b): I promise to give you b.

$$\left. \begin{array}{c} A_1 \xrightarrow{-b} \blacksquare A_2 \\ A_1 \xleftarrow{+b} A_2 \end{array} \right\} \quad (3.41)$$

In the final example. it is tempting to think that there is an equivalence between a promise to accept and an imposition to give, because if an agent imposes 'give me b', it seems to be willing to receive it. However, this is not necessarily true.

Lemma 1 (Inequivalence of − promise with + imposition). *Impositions on an agent to give to oneself do not imply a willingness to accept the imposition:*

$$A_1 \xrightarrow{+b} \blacksquare A_2 \not\Rightarrow A_2 \xrightarrow{-b} A_1 \quad (3.42)$$

Similarly, promises to receive do not imply imposition, obligation or promise to give:

$$A_1 \xrightarrow{-b} \blacksquare A_2 \not\Rightarrow A_2 \xrightarrow{+b} \blacksquare A_1 \quad (3.43)$$

Such implications would violate the autonomy of the agents.

3.11. TRANSFER OF INTENTIONAL BEHAVIOUR BETWEEN AGENTS 41

The above seems clear in an engineering scenario, as say applied to machinery. There, one can simply say that there should be no implicit suggestion made by a promise to use; however, in a human context an implication to accept might be felt to be an implicit obligation to offer. Where human agents are involved, empathy can complicate the semantics of such signalling.

Example 11. *'I promise to receive you at my wedding' suggests an obligation to attend.*

Example 12. *An online shopping site promises that if you click on a book, they will deliver it to you. This generally does not make shoppers feel that they should make a purchase. However, if a friend says they promise to use any old clothes you can provide, you might feel an obligation to provide some.*

Example 13. *Send me the data (blocked by a firewall promise to deny access).*
Invite a foreign guest (doesn't have a passport visa or promise to allow entry).

Example 14. *Give me the toy, no I won't take it!*

An autonomous world in which there are only impositions cannot be a complete description of intent. These examples show that two autonomous parties cannot avoid promises in the mutual signalling of intent. Impositions can be avoided however, by creating implicit imposition with use-promises.

3.11.6 COMPLEMENTARITY AND ± DUALITY: PUSH VS PULL

Each promise graph, classified in terms of $+$ and $-$ promise types has at least two dual or complementary viewpoints. One many exchange the $\pm$ labels on autonomous agents by redefining the promises and impositions accordingly to transform between active and passive descriptions of behaviour.

Example 15. *Consider a client-service relationship between two agents: a provider A_P such as a shop or service, and a client A_C or customer. In an active view, we may imagine the client imposing on the provider for a service (e.g. by demanding to receive X):*

$$A_C \xrightarrow{-X} A_P \qquad (3.44)$$
$$A_C \xleftarrow{-(-X)} A_P \qquad (3.45)$$

In this version, the client imposes on the provider to accept $(-)$ its order/request for X, which the provider promises $(-)$ to use (and hence honour) the order. This is an active or push *description.*

Either of these promises can be rewritten by altering the body of the promise. The imposition can be relabelled as the imposition of a command to give (+X) the required good or service rather than to receive the order.

$$A_C \xrightarrow{+\text{give me}X} \blacksquare \; A_P \tag{3.46}$$

$$A_C \xleftarrow{-\text{give me}X} A_P \tag{3.47}$$

i.e. I accept your imposition. Similarly, the promise to receive the command can be relabelled as a promise to give (+) the desired good or service.

$$A_C \xrightarrow{+\text{give me}X} \blacksquare \; A_P \tag{3.48}$$

$$A_C \xleftarrow{+X} A_P \tag{3.49}$$

i.e. I am offering you X regardless of your imposition. These perspectives can be functionally indistinguishable, but this depends actually on the discerning capabilities of agents in the scope of the declarations assessing them. Some agents might be able to tell the difference between one version and another, others might not.

By relabelling, we may further transform from an active to a *passive* or *pull* description, free of impositions, in which the provider promises its service and the client promises to use the service:

$$A_C \xrightarrow{-\text{service}} A_P \tag{3.50}$$

$$A_C \xleftarrow{+\text{service}} A_P \tag{3.51}$$

The symmetry between ± promises is a fundamental one, what physicists would call *time reversal symmetry*, in this case 'who goes first'. The assumption is often that one can only use something that has been offered, however readers should be careful not to confuse promises with actions. One can certainly promise to use a service before it has been offered, even if the actual keeping of the promise occurs in a different causal order[21].

Lemma 2 (Complementarity transformations). *For any description of behaviour in terms of promises and impositions there exists at least one transformation of the bodies and types, the form*

$$A \xrightarrow{+b_1} \blacksquare \; B \;\leftrightarrow\; A \xrightarrow{-f_1(b_1)} \blacksquare \; B \tag{3.52}$$

$$A \xrightarrow{-b_2} \blacksquare \; B \;\leftrightarrow\; A \xrightarrow{+f_2(b_2)} \blacksquare \; B \tag{3.53}$$

$$A \xrightarrow{+b_3} B \;\leftrightarrow\; A \xrightarrow{-f_3(b_3)} B \tag{3.54}$$

$$A \xrightarrow{-b_4} B \;\leftrightarrow\; A \xrightarrow{+f_4(b_4)} B \tag{3.55}$$

which describes behaviour that is indistinguishable to an observer.

3.12. PARTICULAR KINDS OF PROMISES 43

We shall return to this topic again after introducing the concept of conditional promises (see section 6.2.2), where further transformation of promises can be made to generate equivalent bundles.

3.11.7 THE EFFECTIVE ACTION OF A COOPERATIVE BINDING

When agents' promises bind, action or influence can propagate in the direction of the promises. A promise binding defines a voluntary constraint on agents. The perceived strength of that binding is an individual value judgement made by each individual agent in scope of the promises.

Consider an exchange of promised behaviour, in which one agent offers an amount b_1 of something, and the recipient promises in return to accept an amount b_2 of the promised offer.

$$\pi_1 : A_1 \xrightarrow[\sigma_1]{+b_1} A_2 \qquad (3.56)$$

$$\pi_2 : A_2 \xrightarrow[\sigma_2]{-b_2} A_1 \qquad (3.57)$$

Then:

- Any agent in scope σ_1 of promise π_1, will perceive that the level of promised cooperation between A_1 and A_2 is likely b_1.

- Any agent in scope σ_2 of promise π_2, will perceive that the level of promised cooperation between A_1 and A_2 is likely b_2.

- Any agent in scope $\sigma_1 \cap \sigma_2$ of both promises π_1 and π_2, will perceive that the level of promised cooperation between A_1 and A_2 is likely $b_1 \cap b_2$.

If an agent offers b_1 and another agent accepts b_2, the possible overlap $b_1 \cap b_2$ is called the effective action of the promise.

3.12 PARTICULAR KINDS OF PROMISES

Certain kinds of promises stand out as special cases and we shall need to refer to these concepts in daily usage. We define some of these here for reference.

3.12.1 COMMITMENTS

The term *commitment* is widely used in a human context. We define a commitment simply to be a promise to which one is committed. In Searle's theory of speech acts, the commissive speech acts are commitments[Sea69], and he seems to take a similar view that not all promises need be commitments. This begs the question, what is a promise to which one is not committed? In fact many promises in the human realm are little more than indications of possible behaviour, based on little consideration of whether they can be kept. A possible interpretation of this is: a promise that one promises not to withdraw.

A commitment is thus a kind of ranking of a promise by the promiser that indicates some effort has been put into estimating the likely outcome of the promise. Since commitments only appear once one reaches a level of agent sophistication in which intentionality plays a role[Sea83] (it doesn't make a lot of sense to speak of commitments for atoms or materials, for instance), we will not spend much time on them in this introductory volume.

3.12.2 Exact and inexact promises

Definition 8 (Exact and inexact promises). *A promise is exact if it allows no residual degrees of freedom, else it is inexact.*

A promise $A_1 \xrightarrow{b} A_2$ is inexact if the constraint $\chi(b)$ has residual degrees of freedom. i.e. if it is not a complete and unambiguous behavioural specification.

Example 16. *A promise body constraining a variable q such that $q = 5$ is an exact promise specification, while $1 < q < 5$ is inexact. The same principle applies to the possible outcome of a promise, however the actual outcome is naturally exact in each measurement.*

Example 17. *When electronic components promise their capabilities within certain tolerances, these are inexact promises, e.g. a resistor that promises a rating of 100 Ohms, plus or minus 5 percent.*

3.12.3 Superfluous or empty promises

Several kinds of promise body serve no purpose. This include promises about outcomes that are inevitable.

Example 18. *A promise that the promiser exists would be necessarily kept. A promise that all recipients will receive a promise would be inevitable.*

A case of this is clearly provided by empty promise.

Definition 9 (Empty promise). *An empty promise is one whose promise body contains no type or constraint of any kind.*

Example 19. *'I promise something or other.'*

3.12. PARTICULAR KINDS OF PROMISES

Empty promises can be assumed to be kept at all times, since there is nothing to verify. Note that an empty promise is not the same as making a promise that something else should be empty, or that something should be zero, since both those cases make a constraint on the content of the 'something'.

3.12.4 COORDINATION PROMISES

In addition to these two fundamental types, it is possible to define other promise patterns with special meanings, such as coordination which we explain below.

- $C(\pi)$ a promise to coordinate with another agent

- $\pm\mathbf{def}(\pi)$ a promise to reveal or heed knowledge about a promise π

Readers should not confuse the body describing the communication of a promise with making of the promise itself, as these are two different kinds of object.

3.12.5 LIES AND DECEPTIONS

In a deception (see also section 2.12), there is always a source and always a target and the target cannot be the same as the source, as an agent cannot (intentionally) deceive itself. Furthermore, we maintain that the target of a deception must be in scope, so there must be a physical documentation and hence a deception necessarily involves a promise and not merely an intention.

Definition 10 (Deception). *A deception consists of two intentions: a documented intention (i.e. a promise) and a non-documented intention, which are incompatible.*

From the definition, we can begin to formalize the notion of a lie or deception. A lie, or deception is a promise made about something the agent knows it cannot accomplish, or that it has no intention of keeping.

- These are promises not respecting the default assumption, since default promises must be kept.

- Only the agent making the lie is capable of detecting it in general.

- A liar could try to avoid being caught by saying "I promise X if I can", but ....this is simply a superfluous promise, equivalent to an empty promise, since the conditional is left dangling. It is an evasion.

A set-formulaic way of representing a deception might be the following. A deception is asserting a promise that it does not intend to keep. i.e. an agent A_1 promises something b that is not necessarily kept to A_2, while telling a possibly third agent C that it will promise something that is both not b and which will be kept.

$$A_1 \xrightarrow{b \notin K} A_2 \qquad (3.58)$$

$$A_2 \xrightarrow{b' \in (\neg b \cap K)} A_3 \qquad (3.59)$$

where $A_3 \neq A_2$, i.e. A_2 does not hear the promise to not do b.

3.12.6 Promises to keep promises

Consider the implications of making promises about promises. Let us define

$$A_1 \xrightarrow{b(\pi)} A_2 \qquad (3.60)$$

to mean a promise π to keep the conditions of the body $b(\pi)$, and let

$$\pi' : A_1 \xrightarrow{\pi} A_2 = A_1 \xrightarrow{A_1 \xrightarrow{b(\pi)} A_2} A_2, \qquad (3.61)$$

mean a promise to keep the promise π. A promise to keep a promise might or might not be as strong as a direct promise. This does not affect the details of expected outcome, only belief in the likelihood of the outcome. Thus we cannot say that (3.60) and (3.61) are the same, but it is reasonable to say that they are equivalent up to a discounting factor $\delta \leq 1$ that affects the belief in the outcome. We may write the notation:

$$\beta(\pi') = \beta(\pi(\pi(b))) = \delta\beta(\pi(b)) \qquad (3.62)$$

and if

$$\pi^{(n)}(b) \equiv \pi(\pi(\ldots \pi(b))), \qquad (3.63)$$

then the possible outcomes are the same, so this is equivalent to a different promise with the same intent

$$\pi^{(n)}(b) \to \pi'(b). \qquad (3.64)$$

however it is somewhat weaker. So we have

$$\beta(\pi^{(n)}(b)) = \delta(n)\beta(\pi(b)) \qquad (3.65)$$

meaning that belief in a promise to keep a promise is weaker by a factor δ than a direct promise. This result may be contrasted with the idempotence or repeated invocation of promises in (3.67) of section 3.13.

3.13 Idempotence in Repetition of Promises

What if we repeat a promise to pay 100 Euros to someone; have we kept our promise if we pay 100 or 200 Euros?

Rule 1 (Separate events have separate types). *Each new labelled promise that is fulfilled by the same type of action, must be labelled by a formally different type promise.*

Certain promise types are therefore related, or we can think of them as being parameterized. For example, the different payments above have different types, otherwise we lose track of total value. e.g. τ_1 is payment 1, τ_2 is payment 2.

Rule 2 (Idempotence of promises). *The promise combination operator is idempotent, so repeating the same promise does not change anything:*

$$\cdot A_1 \xrightarrow{b} A_2, A_1 \xrightarrow{b} A_2 \equiv A_1 \xrightarrow{b} A_2. \tag{3.66}$$

Or in simpler notation:

$$\pi(b) \cdot \pi(b) \equiv \pi(b). \tag{3.67}$$

and if

$$\pi^n(b) \equiv \pi(b) \cdot \pi(b) \cdot \ldots \pi(b), \tag{3.68}$$

then by implication

Lemma 3 (Promise notation is idempotent).

$$\pi^n(b) = \pi(b). \tag{3.69}$$

The repetition of a free-standing promise n times is the same as a single promise.

However, the belief in the outcome of the promise might be strengthened by the repetition of the promise, so we may write the notation:

$$\beta(\pi^n(b)) = \alpha(n)\beta(\pi(b)), \tag{3.70}$$

meaning that belief in the outcome is amplified in the repetition of the promise by some factor $\alpha(n) \geq 1$. This means one must be careful in reducing $\pi^n(b)$ algebraically: this is not allowed when the promise is the subject of an assessment (see section 5.2.2).

3.14 Promises involving groups, anonymous and wildcard agents

We can define notation for a promise from one set of nodes to another set of nodes using set curly-braces:

$$\{A_i\} \xrightarrow{b} \{A_j\} \qquad (3.71)$$

i.e. the set of agents $\{A_i\}$ promises b to another set $\{A_j\}$ of agents.

We can now suppose what happens when referring to third-party agents in the body of a promise. In some circumstances, this can apparently lead to contradictions. As a general rule, it seems to be in the interest of clarity to avoid this where possible.

The general promise notation above is designed to avoid the problem of referring to other agents in the body of a promise, but to make some promises we need to talk about third parties. Let us denote:

- Unspecified agents "$A_?$"

- Anonymous agents "A_{anon}".

Ensembles of these can be denoted with braces, e.g. $\{U\}$ would be an ensemble of unspecified agents. We can write

$$A_{\text{anon}}(\{A\}_1) \xrightarrow{b} A_{\text{anon}}(\{A\}_2) \qquad (3.72)$$

to mean one anonymous agent from the ensemble $\{A\}_1$ makes a promise to another anonymous agent from the ensemble $\{A\}_2$.

Example 20. *Usage of anonymous and unspecified agents.*

1. $(A_{\text{anon}}) \xrightarrow{b} (A_?)$: *someone promises (anonymously) that b will happen to someone. For example, a death threat.*

    Terrorist$[A_?] \xrightarrow{\text{attack}}$ President$[A_?]$: *Terrorist promises President that one of his operatives will attack someone.*

    Terrorist$[A_{\text{anon}}] \xrightarrow{\text{attack}}$ President$[A_?]$: *Terrorist promises President that an anonymous individual will attack someone.*

2. Let T be a set of trusted parties, then

$$\text{Parent}[A_?(\{T\})] \xrightarrow{\text{collect from school}} \text{Child} \qquad (3.73)$$

    means, a parent promises its child that one of the trusted set will collect the child from school.

3.15. GOALS OR RANKING OF INTENTIONS

3. Let T be a set of trusted parties, then

$$\text{Child}[A_{\text{anon}}] \xrightarrow{\neg U(\text{collect from school})} \text{Parent} \qquad (3.74)$$

The child promises its parent that it will not accept lifts from strangers.

$$\text{Child}[A_?(-\{T\})] \xrightarrow{\neg U(\text{collect from school})} \text{Parent} \qquad (3.75)$$

The child promises its parent that it will not accept lifts from someone who is not in the list of authorized minders..

3.15 GOALS OR RANKING OF INTENTIONS

When agents work together, one often speaks about the goal of their activities, implying an alignment of purpose. The term goal is not significantly different from an 'intention', indeed there is no reason to inflate the terminology with a new concept. Just as commitments are merely promises that an agent has singled out as being deserving of a special name due to a sense of priority, so goals are special intentions that may or may not have been made public. When agents work together, they might or might not share the same goals, even if they make the same promises.

Definition 11 (Goal). *A goal is an intention with special significance to the agent that maintains is. It is often used as an umbrella intent for a bundle of promises.*

The idea of a goal brings back the issue of value judgement, or how an agent singles out particular promises as being special. We return to this issue more fully in chapter 9. In short, if a goal is met, i.e. the intention was realized with a positive outcome, then it can be said to yield positive value to an agent. There is thus a simple numerical or economic approach to the identification of special intentions.

CHAPTER 4

COMBINING PROMISES INTO PATTERNS

The usefulness of promise concepts lies largely in their properties as 'atomic' or elementary parts of a system. However, it is the way we build larger structures by combining them into *systemic designs*, which accounts for the usefulness of a theory. This chapter describes a few such patterns.

4.1 BUNDLES OF PROMISES

The simplest structure is to collect promises into a kind of logical container with a name. We call this a promise *bundle*. Promise bundles form the basis for understanding system *components* and the functional *roles* they play. They can be defined as absolute quantities, or as parameterized patterns to be re-used as templates for expressing variation.

A basic promise bundle is any collection of promises between any set of agents that we choose to group into an entity.

Definition 12 (Promise bundle). *Let $S, R \subseteq A$ be arbitrary sets of agents, possibly overlapping. A promise bundle is any collection of promises made from agents in S onto agents in R. We denote this*

$$S \stackrel{B}{\Longrightarrow} R \equiv \{s_1 \stackrel{b_1}{\longrightarrow} r_1, \ldots, s_1 \stackrel{b_n}{\longrightarrow} r_n\}. \tag{4.1}$$

B is the name of the bundle, and represents the collective content of the member promises.

4.2. PARAMETERIZED BUNDLES

Example 21. *We might choose to collect all our promises about driving a car into a driving bundle, and all of our promises about healthy eating into a diet bundle.*

Bundles are logical containers, and they may overlap as sets and collections may. The same promise can be part of multiple bundles. The main purpose of bundles is to offer a way of assigning names to simple functional patterns.

Definition 13 (Homogeneous promise bundle). *Let $S, R \subseteq A$ be arbitrary subsets of equal dimension d, from the set of agents. A promise bundle is any collection of promises made (one-to-one) from agents in S onto agents in R. We denote this*

$$S \xRightarrow{B} R \equiv s_1 \xrightarrow{b_1} r_1, \ldots s_d \xrightarrow{b_d} r_d. \qquad (4.2)$$

where B denotes the collection of promise bodies $b_1, b_2, \ldots, b_d$.

Promise bundles are important when defining the parallel composition of promises into patterns, especially for the purpose of discussing system organization and componentization.

4.2 PARAMETERIZED BUNDLES

A parameterized bundle is a pattern like a template, made up of not-fully-specified or conditional promises, in which some of the conditions to be fulfilled are yet to be defined, as promises to the use or accept information from external agents. In a computational sense, parameterized patterns are like functions with arguments. We plug in certain values in order to re-use a generic pattern. The value returned by the function could be interpreted as the collective state of promises kept and not kept.

Definition 14 (Parameterized promise bundle). *Let B be the collection of promises with bodies $b_1, b_2, \ldots, b_d$ belonging to a promise bundle. The bundle is said to be parameterized if at least one of the bodies b_i represents a promise to use information from an external agent, which is subsequently used to specify the choices made in the other promises.*

Example 22. *A website might promise to deliver online goods, but you have to promise to tell them which ones. The bundle of promises the website would keep might be basically the same for any choice, but would still depend on the exact choice input by the customer. Thus the bundle would include a promise to use information from the customer, before dispatching anything.*

Example 23. *In the computer language CFEngine, a parameterize bundle is written as in the example below. This bundle consists, for illustration only, of a single promise whose body consists of statements of the form* attribute $\Rightarrow$ value. *The parameter values are substituted by writing* $ (name) *:*

```
bundle myname(parameter1, parameter2)
{
files:                                  # The promise type

    "$(parameter1)"                     #  The promiser

        create => "true",               # The body
       edit_line => "$(parameter2)";    # $ represents substition
}
```

The bundle has a name 'myname' and two parameters corresponding to use-promises. When this bundle is promised one promises values for these parameters at the same time:

```
"promiser"
   usebundle => myname("file1", "content of file");
```

Parameterized bundles are an important pattern for the practical use of promises, as they allow us to make generic 'forms' that compress the total amount of work into smaller, reproducible promise templates. Many standard contracts have this property, such as application forms, etc.

4.3 Promise valency and saturation

To develop the formal chemistry of intent, we need to clarify how many agents a promise can support. In other words, how many 'slots', 'binding sites' or occupyable resources does an agent have, to share between promisees?

Declaring a finite number of such slots, explicitly allows for a simple discussion of resource exclusivity around promises. The concept is basically analogous to the valences (oxidation numbers) of electrons in physical chemistry. Think also of the binding sites for receptors, viruses and major histocompatability proteins in biology.

4.3. PROMISE VALENCY AND SATURATION

Definition 15 (Valence of an agent promise, and overcommitting). *A promise which provides $+b$ to a number of other agents may specify how many agents n for which the promise will be kept exclusively. The valency number of an exclusive promise is a positive integer n, written*

$$A \xrightarrow{+b^n} \{A_1, \ldots A_p\}. \tag{4.3}$$

A promise body may be called over-promised (or over-committed) if $p > n$.

The concept of valency for promise bindings was introduced in [Bur15], where the notation $b\#n$ was used for valency. Since that collides with the use of $\#$ for promise incompatibility, we have chosen a superscript notation going forward.

Example 24. *A reserved parking area promises 10 spaces, to 20 employees. The parking promise is over-committed, since it cannot keep all of its promises simultaneously.*

Over-promising is not a problem unless all of the promisees accept the promise, and promise to use it. Thus a separate concept of saturation arises by using up all of the valence slots:

Definition 16 (Use-promise saturation). *Suppose we have*

$$A \xrightarrow{+b^n} \{A_1, \ldots A_p\} \tag{4.4}$$

$$\{A_1, \ldots A_m\} \xrightarrow{-b^m} A \tag{4.5}$$

is saturated if $m \geq n$ concurrently.

It is useful to define a function whose value is the net valence of a particular type of promise body.

Definition 17 (Net valence of a promise graph and utilization). *$\pm b$, for a collection of agents $\{A_i\}$:*

$$\mathbf{Valence}(b; \{A_i\}) = \sum_i \mathbf{Valence}(b; A_i) \tag{4.6}$$

$$= n - m \tag{4.7}$$

Hence we may assign an integer value to the level of usage, or a rational fraction m/n for utilization of the resource. If this fraction exceeds unity, or the net valency is negative, the keeping of the promise effectively becomes a queue of length $|m - n|$, requiring the agent to multiplex its resources in time to keep its promise.

Example 25. *Consider the two agents A_1, A_2:*

$$A_1 \xrightarrow{+b^2} A_2 \qquad (4.8)$$

$$A_2 \xrightarrow{-b^3} A_1 \qquad (4.9)$$

A_1 *offers two possible slots for its promise of $+b$, while A_2 requests three units of it, leaving a net deficit:*

$$\textbf{Valence}(b; A_1, A_2) = -1 \qquad (4.10)$$

This notation allows us to simplify the discussion of occupancy and tenancy in later sections.

Example 26. *Consider the following promises made by a network switching device:*

$$\text{switch} \xrightarrow{+(10Gb)^48} \text{client} \qquad (4.11)$$

$$\text{client} \xrightarrow{+(1Gb)^1} \text{switch.} \qquad (4.12)$$

The switch makes 48 promises offering 10Gb 'bandwidth' to the clients. The client accepts one valency slot (leaving 47 more), and promises to consume only 1Gb of the maximum possible 10Gb.

4.4 AGENT ROLES

Promise bundles may play different roles in a system of promises. One may use bundles to decompose sets of promises into possibly overlapping categories, in order to model scenarios. This is a realistic approach to understanding function in a network of agents[22]. The concept of roles from promises was first discussed in [BFa, BFb, BF06]. In this section we shall try to formalize the notion of a role. This will be key to understanding the composition / decomposition of systems.

4.4.1 FORMAL DEFINITIONS OF KINDS OF ROLE

We can imagine roles from coming from a number of different interpretations of what identifies a subset of promises.

- Groups of agents that send S and receive R promises (i.e. promisers and promisees).

- Grouping agents that give the same $+$ promise (i.e. functional similarity) – role by association.

4.4. AGENT ROLES

- Grouping promisees that are made the same promise (i.e. recipient similarity) – role by appointment.

- Grouping agents that use the same promise (i.e. functional dependency).

- Grouping agents that work together so that a representative can make a promise like a single agent (cooperation).

Roles may thus be associated both with *types* of promises τ, compositions of promises and agents, and the topology of the promise graph itself. We shall meet roles in several other situations, including in the definition of components and organizations.

In this respect, we can associate a collection of agents with a role, given that they make (or use) the same set of promise types. In other words, we are guided by the typed structure of the graph alone.

Definition 18 (Roles by association). *Promisers that* make *the same promise may be grouped by their associated function. Let $\mathcal{A}$ be any set of autonomous agents, divided into senders S and receivers R. The family of sender subsets in a promise graph, which take part in promises with body types $t(b)$, define b-roles by association. Note that the promise made might be a give or receive promise (i.e. of type $\pm$).*

Definition 19 (Roles by appointment). *Promisees that are* made *the same promise may be grouped by their associated function, regardless of whether they themselves promise to use the promise given. Let $\mathcal{A}$ be any set of autonomous agents. The family of receiver subsets $R \subseteq \mathcal{A}$ in a promise graph, which take part in promises with body types $t(b)$, define b-roles by appointment.*

Note that promise roles are identified empirically, not pre-decided by design.

Example 27. *The set of all agent nodes $W \subseteq \mathcal{A}$ that promises to provide web service to any agent $R \in \mathcal{A}$:*

$$W \xrightarrow{\text{web}} (R \in \mathcal{A}) \tag{4.13}$$

plays a role, which we can call 'web server'. This is called a role by association, since the members of W are not connected in any other way than they make the same promise. The set of nodes $C \subseteq \mathcal{A}$ that is promised web service by any agent:

$$(S \in \mathcal{A}) \xrightarrow{\text{web}} C \tag{4.14}$$

plays a role, which we can call 'web client'. This is called a role by appointment, since the members of C are pointed to by a promise of the same type.

Example 28. *In an electronic circuit, resistors and capacitors play roles by association. They are agents that make promises to resist current and store charge, respectively, within specified limits. Resistors with different values might play different roles depending on how specifically the promises are stated.*

Example 29. *Membership clubs, leaders and pop stars are all examples of roles by appointment. Agents (people) promise to follow them and subscribe to their activities, whether they are aware of this or not.*

Definition 20 (Coordinated roles or roles by cooperation). *Let $\mathcal{A}$ be any set of autonomous agents. A subset of agents C that are collectively involved in providing a promise with body types $t(b)$, define b-roles by cooperation. In order to cooperate, all agents in the group supporting the role mutually promise $C(b)$ in a complete graph, i.e. to cooperate in the performance of b. (See fig. 4.2)*

Lemma 4 (Cooperative ensemble). *$\{A_c\}$ is a cooperative ensemble if and only if each agent participates in a mutual agreement to coordinate with every other agent in the promise:*

$$\{A_c\} \xrightarrow{b} A_i \tag{4.15}$$

for some node or set of nodes A_i. Every agent in the group has a promise

$$\{A_c\} \xrightarrow{C(b)} \{A_c\} \tag{4.16}$$

for the given b.

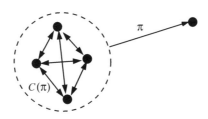

Figure 4.1: A cooperative role is defined when an ensemble of agents, formed through mutual cooperation, agrees to offer the same promise to another agent, or set of nodes.

4.4. AGENT ROLES

4.4.2 GRAPHICAL INTERPRETATION OF ROLES.

Because promises form a graph, or behavioural network, there is a natural topological understanding of what it means for an agent to play a specific role in a system. Agents that make the same patterns of promises, may be said to play the same roles. In technical terms, we are guided by the topology of the typed graph alone[23].

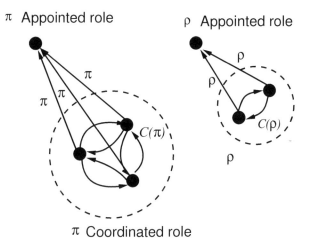

Figure 4.2: Typed roles distinguish between the type of promise used to define the structure. Natural groupings result from the autonomous behaviour.

- An *appointed role* arises when an agent or group is being pointed to by a set of identical promise arrows (role-sink). In a sense, the agents are effectively voted for (like the candidates in an election).
- Roles by *association* are played by uncoordinated agents that happen to make the same promises, such as the voters in an election.
- A *cooperative role* arises when a group of agents is united by making a coordination/subordination promise to one or more other nodes. They take on the role of a 'community' or 'organization' of dependables.

4.4.3 SYMMETRY OF AGENTS UNDER ROLES

Roles are labels for agents or groups of agents derived either from observed or promised behaviour. The roles played by agents in an ensemble can be assigned simply by looking for all repeated patterns of promises. A role is thus a pattern. Since similar promises will lead to similar observed behaviour, we may define role labels for each distinct combination of promises that occurs in a promise graph.

- *Differentiated behaviour:*

  Agents that behave differently, e.g. perhaps partitioned into a division of labour when cooperating, or simple independent.

- *Undifferentiated behaviour:*

  Agents play identical roles in the ensemble and require no specific labels, since all promises are made by each agent.

Undifferentiated behaviour might be coincidental i.e. un-calibrated, like a disordered gaseous phase of matter (all the component elements coincidentally make identical unconditional promises but never interact with one another). Alternatively, agents might have agreed to behave alike through interaction (like in a solid phase of matter).

Normally we are only interested in the possibility of coordinated, collective phenomena, but a phase transition from one to the other is possible and we shall describe this elsewhere[Bal04].

4.5 Some common patterns or roles

Common nomenclature for roles in society and technology motivate the following elementary patterns.

4.5.1 Client or customer

A client is an agent that makes a request (an imposition to accept), with body R, of another agent.

$$\text{Client} \xrightarrow{-R} A_?. \tag{4.17}$$

4.5.2 Consumer or dependant

A consumer is an agent that depends on or promises to use a promise or service S provided by another agent on a long term basis.

$$\text{Client} \xrightarrow{-S} A_?. \tag{4.18}$$

4.5. SOME COMMON PATTERNS OR ROLES

4.5.3 SERVER OR SUPPLIER

A server is an agent that promises to accept requests for a service and return that service S on a long term basis, as in figure 4.3.

$$\text{Server} \xrightarrow{-R,+S} A_? \qquad (4.19)$$

This is the logical counterpart of a client or consumer role.

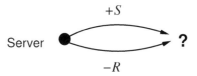

Figure 4.3: The server or supplier role.

4.5.4 SERVER POOL (COORDINATED SERVERS)

A server pool is a collection of peered server Server_i agents, that behave in an identical fashion.

$$\{\text{Server}_i\} \xrightarrow{-R,+S} A_? \qquad (4.20)$$

$$\{\text{Server}_i\} \xrightarrow{C(S),C(-R)} \{\text{Server}_j\} \; (i \neq j) \qquad (4.21)$$

where i, j run over the labels of the different servers.

4.5.5 PEER (VICE-VERSA EXCHANGE)

Peers are agents that mutually mutually consume promises of the same type, as in figure 4.5.

$$\text{Peer}_1 \xrightarrow{+P_1} \text{Peer}_2 \qquad (4.22)$$

$$\text{Peer}_1 \xrightarrow{-P_2} \text{Peer}_2 \qquad (4.23)$$

$$\text{Peer}_1 \xleftarrow{+P_2} \text{Peer}_2 \qquad (4.24)$$

$$\text{Peer}_1 \xleftarrow{-P_1} \text{Peer}_2 \qquad (4.25)$$

Example 30. *Peer 1 shares its knowledge with peer 2 and vice versa. This relationship is common in Internet traffic routing technology.*

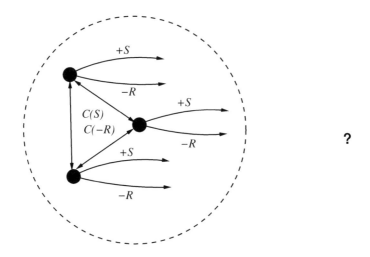

Figure 4.4: The server pool superagent.

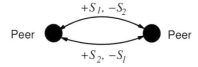

Figure 4.5: The peer role.

4.5.6 QUEUE DISPATCHER OR FORWARDING AGENT

A queue dispatcher is an agent that sits in front of a number of server roles and promises to accept requests, then forwards or dispatches the request to one and only one of the servers, based on some policy (see figure 4.6).

$$\text{Dispatcher} \xrightarrow{-R, -\{Q_i\}} \text{Server}_i \qquad (4.26)$$

$$\text{Dispatcher} \xrightarrow{R} \blacksquare \text{Server}_i \qquad (4.27)$$

$$\{\text{Server}_i\} \xrightarrow{Q_i} \text{Dispatcher}. \qquad (4.28)$$

The dispatcher typically uses promises Q_i of the servers S_i to keep information about the queue lengths or workloads they experience. This enables it to select one of them to impose extra work onto.

This pattern is typically used as an interface to a server pool in order to share the load between the servers pattern[24].

4.5. SOME COMMON PATTERNS OR ROLES

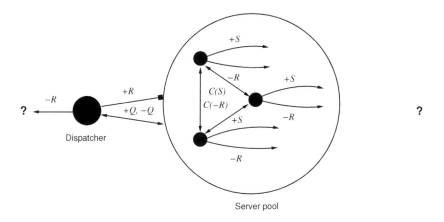

Figure 4.6: The dispatcher role.

4.5.7 TRANSDUCER

A transducer is an agent that promises to accept one kind of promise or request R and, in turn, promises a transformation of that 'input' as 'output', here represented as a function $T(R)$ of R.

$$\text{Transducer} \xrightarrow{-R} A_{?1} \qquad (4.29)$$
$$\text{Transducer} \xrightarrow{+T(R)} A_{?2} \qquad (4.30)$$
$$(4.31)$$

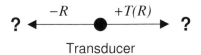

Figure 4.7: The dispatcher role.

Example 31. *An electrical appliance has been designed to conditionally promise functionality if it can use the electrical power. The component was designed in the USA, but, in the UK, it cannot find a power point that promises the required voltage or pin configurations. Simply put, you can't plug the device into the power socket. The conditional promise made was too specific. However, it is possible to buy an adapter which promises to bridge the gap between what is available and what is required.*

$$\text{UK socket} \xrightarrow{square\ 3\ pin\ 240V} A_? \quad (4.32)$$
$$\text{US appliance} \xrightarrow{function|flat\ 2\ pin\ 120V} A_?$$
$$\text{Adapter} \xrightarrow{flat\ 2\ pin\ 120V|square\ 3\ pin\ 240V} \text{US appliance}$$
$$\text{Adapter} \xrightarrow{U(square\ 3\ pin\ 240V)} \text{UK socket} \quad (4.33)$$

4.5.8 THE MATROID HUB CONSTRUCTION (CALIBRATION)

Consider how to represent standardized types of agent, promising point-like properties. If all agents are independent, then how do all people follow the same rules? How do all electrons promise the same charge? How do all the chocolate bars taste the same by promising the same recipe?

There are two possible representations of a collective standardized promise: by gradual consensus, and by following a leader. Let's suppose that agent A_1 promise to its colour is blue.

- The first case is to establish standard by *equilibration* (see figure 4.8):

$$A_1 \xrightarrow{+\text{I am Blue}} A_2 \quad (4.34)$$
$$A_2 \xrightarrow{-\text{I am Blue}} A_1 \quad (4.35)$$

A_1 merely asserts a property of itself to an observer A_2. The 'observer' A_2 can take or leave the promise of blueness, thus it must promise to use or accept the assertion, (it might be colour blind, for instance, and hence effectively not promise to use the information). Notice here that the conceptual world of blueness lives entirely within the body of the promise, and does not affect the type of objects between which promises are made. This means that the agents themselves have to have sufficient capabilities to discriminate blueness. Because the agents are independent, any agent might make the same promise with a different standard, and a priori there is be no objective calibrated standard for blueness. Different agents might interpret this promise differently too. This is why the collective has to come to an agreement about the meaning of blue.

Multiple agents may now in two ways: either by coordinating their definitions individually in a peer-to-peer clique:

$$\{A_i\} \xrightarrow{\pm\text{Blue},C(\text{Blue})} \{A_j\}, \forall i,j \quad (4.36)$$

- The second representation is called the *matroid construction*, and operates like a set of 'standard candles'. We introduce a special agent (the 'blueness' agent)

4.5. SOME COMMON PATTERNS OR ROLES

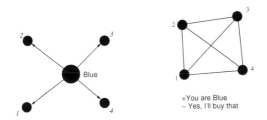

Figure 4.8: Global symmetries may be enabled by calibrating a property, either by equilibrating with a single source or with all the neighbouring peers.

which is the custodian of the named property—a hub of 'truth', like a Trusted Third Party.

$$A_{\text{Blue}} \xrightarrow{+\text{Blue}} A_1 \qquad (4.37)$$

$$A_1 \xrightarrow{-\text{Blue}} A_{\text{Blue}} \qquad (4.38)$$

This is a simple subordination to an authority. For multiple agents:

$$A_{\text{Blue}} \xrightarrow{+\text{Blue}} \{A_i\} \qquad (4.39)$$

$$\{A_i\} \xrightarrow{-\text{Blue}} A_{\text{Blue}} \qquad (4.40)$$

which then acts as a kind of hub for the property of blueness (see fig. 4.8).

The special agent, whose function it is to label things as blue, now promises this quality as if providing the property as a single point of service.' The clients of this property needn't have the capability to discriminate blueness, as they only have to accept the property on trust, by association with the hub. Association with this source of blueness is what gives A_1 the property. To be blue, all A_1 has to do is to promise to use the service. It can subsequently promise its own blueness conditionally:

$$A_1 \xrightarrow{+\text{I am blue} \mid \text{Blue}} \qquad (4.41)$$

The scalar promises bear an obvious resemblance to the use of 'tags' or 'keywords' to label information documents and inventory items in databases. They act as orthogonal dimensions to the matroid that spans agents in the ordered phase of a spacetime.

Notice that, since each property has at least one single vertex associated uniquely with it (the source for that property), the set of links emanating from it is automatically an *independent set*. Each property or type of promise therefore refers to an intrinsic quality and is, in fact, a *basis* 'vector' (see chapter 16), belonging to a matroid (see fig. 4.9). An agent that promises a mixture of properties can promise weighted amounts of these standard candles. What is interesting is that, unlike a vector space, this vector ends at a singularity, like a charge radiating lines of force.

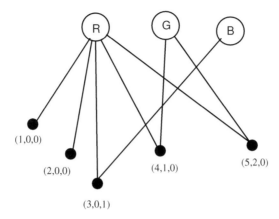

Figure 4.9: Matroid basis for global properties with three property hubs adding three components to the coordinate tuples. The hub has an appointed role, and the appointees inherit an understanding of its calibrated standard promise.

Example 32 (RGB pixels). *The pixels on a television or computer screen are formed from mixtures of three colours: red, green, and blue. Each pixel promises that its red, green, and blue colours are the same as its neighbours, so that the whole screen is colour calibrated. As the pixels change their mixtures of these three properties the effective colour of the screen changes, like a quasi-vector space.*

Example 33 (Country or brand?). *How do we define a country? What is the binding force that keeps it together? Consider a country, which starts from some initial cohesive territory. The country expands and conquers disconnected islands and territories around the world, forming an empire. Over time, these overseas realms are granted autonomy, but the reputation of the country persists, as well as a common passport. Should we consider it still a country, a membership club, or just a brand identity?*

Chapter 5

Assessment and Measurement

We must now consider how to verify whether a promise has actually been kept or not. The role of observation is a critical one in determining what agents believe about the worlds they occupy. Promises span every possible kind of behaviour, thus we choose a form of assessment that is generic, reducing every assessment to a binary determination of whether a reference promise has been 'kept' or 'not kept'. This is simplistic, but has desirable properties.

Assessment of promises is an inherently subjective matter. Each agent may arrive at a different assessment, based on its observations ('I am (not) satisfied with the outcome of events). Assessments must therefore respect the relativity of agent perspectives.

5.1 Elements of promises and assessments

The key elements in the making of promises are summarized by table 5.1.

Component	Description
Promise	What outcome we intend
Algorithm	How a promise is kept
Motive	Why a promise is made
Context	Where, when a promise is made (patterns)

Table 5.1: The context of a promise

Similarly, the corresponding elements of assessing whether a promise has been kept may be written as in table 5.2. While not necessarily required or supplied, the autonomy of agents allows us to separate these matters, and each of them warrants an explanation for a full documentation of intent.

Component	Description
Outcome	What was measured during the assessment?
Context	Where was the sample taken?
Algorithm	How was the sample taken?
Motive	Why the assessment was made
Expectation	What were we expecting to find?

Table 5.2: The context of an assessment

5.2 Defining assessment

Every promise may be assumed assessable. Indeed, from the time a promise is made, a process of assessment begins in all agents that are in scope. This assessment may or may not be rational, but it is part of an on-going estimation[25].

Example 34. *Human agents often make assessments based on no evidence at all. Trust and even prejudice can form the basis for expectation. One might say: 'when I release the hammer, it falls. This has always happened before, I see no reason why it would not happen now'. In a familiar context, this assessment might be reliable; in outer space, it would be false. Whether inference is useful or not depends on a separate assessment of its relevance.*

5.2.1 A definition of promise assessment

Let us define a set of impressions I to mean a set of sampled observations resulting from promises to accept (-) from a number of sources.

Definition 21 (Assessment). *A 'decision' by a single agent O about whether a promise π has been kept or not, usually written $\alpha_O(\pi)$ for an assessment by O about π.*

An assessment by agent A of a promise π may be any function of a set of impressions I, that maps to the promise being 'kept' or 'not kept' at a certain time. It is written $\alpha_A(\pi; t_i, t_f; I)$, where t_i is the initial time and t_f is the final time of the assessment interval, and I is a set of impressions on which the judgement is based.

5.2. DEFINING ASSESSMENT

Note that the assessment function itself makes a promise, namely one to supply its determination of the outcome, thus it is not strictly a new kind of object in the theory of promises. The impressions I are to be thought of as points of data, either invented, received, or specifically measured about the promise. These might include evidence from measurements, or even hearsay from other agents.

Lemma 5 (Assessments are relativistic non-invariants). *Assessments are a result of contextual observation, and therefore may be unrepeatable and particular to each agent at each moment in time.*

Observations by agents at different places and times essentially have the status of random variables.

Example 35. *Two agents, Alice and Bob, order a pizza. The menu promises a spicy pizza, which they accept. Alice assesses the pizza to be spicy, Bob assesses the pizza to be not spicy.*

Example 36. *According to Newton's Laws, in physics, the energy to launch a rocket into orbit is X Joules, the difference in potential energy between the ground and orbit. An engineer works out the fuel required to launch the rocket, but it falls short. Newton's laws did not account for the friction in the atmosphere. Did Newton's Laws keep their promise?*

5.2.2 Auxiliary definitions of assessment

It is useful to define two functions of a promise that indicate assessments of outcome. These correspond roughly to the two interpretations of statistical data: namely a belief (Bayesian) interpretation, and a frequentist or evidential interpretation.

Definition 22 (Belief in outcome $\beta(\pi, t_i, t_f, I)$). *An assessment made without direct observation $\beta : \pi, t_i, t_f \to [0, 1]$ of the likelihood that a promise π will be kept within a stated time frame $t_f - t_i$.*

Definition 23 (Evidence of outcome $\epsilon(\pi, t_i, t_f, E)$). *An assessment made with partial information $\epsilon : \pi, t_i, t_f \to [0, 1]$ of the likelihood that a promise π was kept within a stated time frame $t_f - t_i$.*

When making assessments of promises, one must be careful in applying algebraic reductions using their idempotence (see note in section 3.13).

5.2.3 TRUE, FALSE AND INDETERMINATE

Three important assessments of an assertion X are written as follows:

$T(X)$	The condition X has been assessed to be true.
$F(X)$	The condition X has been assessed to be false.
$?(X)$	The condition X is unknown or undecidable.
$\Pr(X) = \beta$	The truth of X is believed to be probably $\beta \in [0, 1]$.

Assessments are sampled at a localized moment in time, i.e. they are events whose condition may or may not persist. Even when a condition has been assessed true or false, the promise of this assessment has limited trustworthiness, as the system might have changed immediately after the assessment was made.

5.2.4 INFERRED PROMISES: EMERGENT BEHAVIOUR

Sometimes systems appear to an observer to act as though they keep certain promises, when in fact no such promises have been made. The observer might simply be out of the scope of the promise, but can still observe its repercussions; or, in fact, no such promise might have been made at all.

Lemma 6 (Only positive information). *Agents only have information about the existence of promises for which they are in scope. They cannot know if there exist promises for which they are out of scope.*

Does this matter? From the perspective of the observer, it makes no difference whether a promise has actually been made or not as long as the agent appears to be behaving as though one has been made. It is entirely within the observer's rights to postulate a model for the behaviour in terms of hypothetical promises.

Example 37. *Science approaches the 'laws' of nature: it appears that the world makes certain behavioural promises, which we can codify into laws. In truth, of course, no such laws have been passed by any legal entity. Nature seems to keep these promises, but no such promises are evident or published by nature in a form that allows us to say that they have been made.*

Based on its own world, with incomplete information, an agent is free to postulate promised behaviour in other agents as a model of their behaviour. This hypothesis can also be assessed in the manner of a promise to self.

Definition 24 (Inferred promise). *From any assessment $\alpha_A(\pi_?; t_i, t_f; I)$ of impressions I, over an interval of time $t_f - t_i$, an agent A may infer the existence of one or more promises $\pi_?$ that fit its assessment.*

5.3. BOUNDARY CONDITIONS FOR ASSESSMENTS

The observer A cannot know whether its hypothesis is correct, even if an explanation is promised by the observed agent, it can only accumulate evidence to support the hypothesis.

Example 38. *Suppose a vending machine is observed to give out a chocolate bar if it receives a coin of a certain weight. Since most coins have standard weights and sizes, a thief who was not in possession of this knowledge might hypothesize that the machine actually promised to release a chocolate bar on receiving an object of a certain size. Without further evidence or information, the thief is not able to distinguish between a promise to accept a certain weight and a promise to accept a certain size, and so he might then attempt to feed objects of the right size into the machine to obtain the chocolate bar. Based on new evidence, he might alter his hypothesis to consider weight. In truth, both hypotheses might be wrong. The machine might in fact internally promise to analyze the metal composition of the coin along with its size and other features.*

5.3 BOUNDARY CONDITIONS FOR ASSESSMENTS

Several possible models exist for assessment, reflecting which points of knowledge an observer has chosen to trust. Such points are referred to as boundary values, or initial values[26].

An assessment may or may not start with an existing promise, i.e. with expectations of some event in the beginning of an experiment. Later, there is a state in which the assessment has been made and the result has been obtained. In between, there are possibly events. There are incoming states and outgoing states, and the transition function between them represents an assessment of the belief in a transition from one to the other[27].

- Retarded assessment: One starts from an initial state; events occur relative to promises made, and an assessment of the final state is made at a later time.

- Advanced assessment: One starts with a knowledge of the outcome state, and infers from the known promise, an assessment or inference of the initial state.

5.4 THE OBSERVATION INTERACTION: MEASUREMENT

A successful measurement is an interaction between two parties, requiring mutual promises. The autonomy of agents implies that the amount of information transmitted is the overlap between the 'willingness to transmit' and the 'willingness to receive' of the agents involved. The 'willingness to receive' acts as a filter on what will be accepted by the receiver[28].

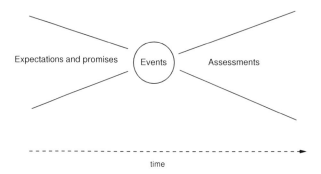

Figure 5.1: One possible model of causal boundary conditions on the assessment process. Here the promise comes first, something happens, and the result is assessed.

$$A_s \xrightarrow{+\text{data}} A_r \quad (5.1)$$
$$A_r \xrightarrow{-\text{data}} A_s \quad (5.2)$$

for sender A_s and receiver A_r. Schematically the resulting promise is the overlap, or intersection between these constraints (see section 3.11.7).

$$\text{Measurement} = \text{Receiver reception} \cap \text{Sender data transmission} \quad (5.3)$$
$$\text{Measurement} = \text{Filtered senses} \cap \text{Observables} \quad (5.4)$$
$$\text{Measurement} = (\text{Capability} \cap \text{Accepted}) \cap (\text{Data} \cap \text{Revealed}) \quad (5.5)$$

Measurement and observation are central to the discussion of system behaviour. The concepts of calibration and coordination must also feature. Promises limit the possible observables in Promise Theory. No information may be exchanged without both promises to be observable and to observe in a binding relationship.

Example 39. *The location and nature of an agent can only be observed if the agent promises to make its position and attributes visible. Camouflage is a deception. Reflectors and bright colours represent promises to be seen in the dark.*

The concept of events plays a major role in the language used in nearly all observable processes. Let's define it here in a way that respects information theoretic transfer. Information is only 'arrives' somewhere when it is sampled.

Definition 25 (Event). *A discrete unit of process in which an atomic change is observed or sampled.*

Notice that it is the promise to accept information that generates an event for the observer, not the promise to offer the information in the first place. The reason for that should be clear: each agent generates its own sense of time according to its own clock and its own private internal information.

We often imagine processes being driven by a flow of events, like a stream[29]. Again, in terms of sampling, this amounts to the following:

> **Definition 26** (Event or message driven agent). *An Event Driven Agent R makes a promise conditionally on the sampling of message events M from a S, with an average rate λ:*
> $$R \xrightarrow{+E|M} \blacksquare\, O \tag{5.6}$$
> *i.e. R can promise an observer O that it acknowledges an event E on receipt of a message M. By Promise Theory axioms, this assumes the prior promises:*
> $$S \xrightarrow{+M|\lambda} \blacksquare\, R \quad \text{or} \quad S \xrightarrow{+M|\lambda} R \tag{5.7}$$
> $$R \xrightarrow{-M|\mu} S \tag{5.8}$$
> *where μ is the queueing service rate.*

Notice that by using the term sampling here, we do not take a position on whether messages were imposed by pushing from S to R, or whether R reached out to S to pull the data. These distinctions are irrelevant to the causal link that results from the message policy. Data are not received until they are sampled by the receiver. Note that there is no timescale implied by the conditional promise in (5.6)—the definition of 'immediate' or 'delayed' response is an assessment to be made by the observer O.

5.5 ALGEBRA OF MEASUREMENT

The need for assessments to be calibrated according to a standard scale is automatically accomplished for each individual agent in its own world. Observational assessments promised by other agents cannot be automatically trusted however as they come with potentially different standards. This suggests a basic algebra of measurement because it implies that all comparisons must be made through a single adjudicator.

Example 40. *Time synchronization across different clocks is a difficult problem because each clock promises to tell the time from its own perspective, but time is relative to an agent's individual world. Synchronization of clocks requires considerable effort and exchanges of information and may not persist. To determine whether clocks agree, a* *third agent could ask A_1 and A_2 to promise their clock times, accept these promises equally, and assess whether their times were equal according to its own world view.*

Calibration may be achieved through cooperation. Recall the shorthand notation $C(b)$ for a pattern of promises that leads to the effective promise "exhibit the same as" (see sections 3.12.4 and 6.3). The coordination promise is used for this (see fig. 5.2). The reduction rule for coordination promises for the case in which A_1 promises A_2 that

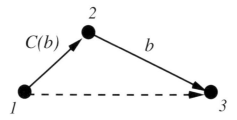

Figure 5.2: Serial composition of a promise and a coordination promise. The dashed arrow is implied by the $C(b)$ promise.

it will coordinate on the matter of b, given that A_2 promises A_3 b follows.

$$\underbrace{A_1 \xrightarrow{C(b)} A_2}_{\text{'Coordinate with'}} , \underbrace{A_2 \xrightarrow{b} A_3}_{\text{Promise}} \gg A_1 \xrightarrow{b} A_3 . \tag{5.9}$$

The coordination promise is transitive.

$$A_1 \xrightarrow{C(b)} A_2 , A_2 \xrightarrow{C(b)} A_3 \gg A_1 \xrightarrow{C(b)} A_3. \tag{5.10}$$

We use this below in the identification of observable properties, since it implies a basis for A_3 to compare A_1 and A_2.

5.5.1 DISTINGUISHABILITY AND EQUIVALENCE OF OUTCOMES

Agents may or may not be able to tell the difference between promised scenarios (see figure 5.3). If an agent does not possess complete information about all promises within a certain scope, it will be unable to distinguish different behaviours, and must consider them to be equivalent.

Example 41. *Consider the scenarios in fig 5.3. The coordination rewriting rule can be applied both forwards and backwards. Since the agent that observes the promises b has no knowledge of what might lie behind it, it cannot tell the difference between these promises by observing the behaviours.*

5.5. ALGEBRA OF MEASUREMENT

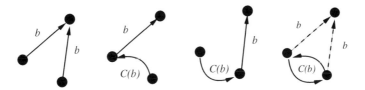

Figure 5.3: Observational indistinguishability leads equivalence of these behavioural configurations or outcomes.

Definition 27 (Equivalence under observation). *Two patterns of promises may be considered equivalent under observation by a third agent if the observing agent cannot distinguish between its assessments of their behaviour.*

Notice that this definition of equivalence is relative to a specific mode of assessment, as indeed it must be. A simple corollary to this is that agents may or may not be distinguishable to other agents, if they promise no properties by which to identify themselves uniquely. Even if agents do promise unique qualities, an observer may not be able to perceive the difference.

Definition 28 (Distinguishability). *An observer O can distinguish two agents A and A', if and only if both agents promise names (scalar attributes),*

$$A \xrightarrow{+name=X} O \qquad (5.11)$$
$$A' \xrightarrow{+name=Y} O \qquad (5.12)$$
$$O \xrightarrow{-name=X} A \qquad (5.13)$$
$$O \xrightarrow{-name=Y} A' \qquad (5.14)$$

that are assessed to be unequal by the observer (i.e. may be promised conditionally to an unspecified agent $A_?$):

$$O \xrightarrow{X \neq Y \mid X,Y} A_?. \qquad (5.15)$$

Distinctions can be observed within the same sample (spacelike comparison) or across subsequent sequential samples (timelike comparison) to detect change. The fact that promises are not always kept means that Observability cannot be treated as deterministic either. Observability requires both promises to be present and kept in order for information to be exchanged. Notice that, in physics we normally attribute properties like distinguishability of particles as bulk properties for bosons or fermions, etc. In Promise Theory, this is a property of the interaction between each pair of agents, not a global property that can be assumed for all observers.

In summary, two promises can be considered equivalent if they have the same assessments.

$$\alpha(A_1 \xrightarrow{S} A_2) \ggg \alpha \left(\left\{ \begin{array}{c} A_1 \xrightarrow{S|X} A_2 \\ A_1 \xrightarrow{\pm X} A_2 \end{array} \right. \right) \tag{5.16}$$

It never makes sense to say that two promises of different types are equivalent, since the outcomes would be different even if the probability of their being kept were coincidentally the same.

5.5.2 Rewriting for cooperative behaviour

Based on the equivalences shown in fig. 5.3 a number of rewriting rules can be formulated expressing equivalences under our observation as modellers with complete information about all agents (having a globally privileged insight that no single agent has). We shall describe them pictorially here to avoid unnecessary formalism. From the symmetry of indistinguishable agents, the implicit promises between the agents must appear in both directions.

1. *Inferring implicit cooperation.* Two agents that behave in the same way to a real or fictitious external observer over some defined timescale can be assumed to be coordinated, and we may introduce symmetrical coordination promises.

2. A corollary to this is that when all agents in an ensemble make identical promises to each other in a complete graph, we can add mutual coordination promises between all pairs. This is easily justified as it reflects the observation that when a number of agents behaves similarly with no labels that otherwise distinguish special roles, an external observer can only say that all of the agents are behaving in a coordinated way. Thus the observer sees a coordinated group for all intents and purposes, although it was not formally agreed by the agents.

3. *Inferring observational indistinguishability.* Any two agents that mutually coordinate their behaviour may be considered to behave analogously over the sampling interval to a hypothetical external observer. This can be formulated by introducing a fictitious promise to the fictitious observer[30].

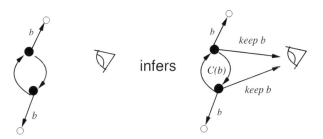

5.6 EMERGENT BEHAVIOUR DEFINED

The idea of inferred promises and observed behaviour allows us to now write a simple definition of emergent behaviour, often referred to in the literature of complex adaptive systems and swarm intelligence[BDT99, Joh01, Hol98, BF07a, BF07b].

Definition 29 (Emergent behaviour). *Behaviour assessed by an observer as consistent with keeping an inferred promise, especially a cooperative promise.*

A positive aspect of this interpretation is that it separates perception of order and speculations about mechanisms from the observational aspects of the phenomenon. With this definition, emergent behaviour can be a property of one or many agents without prejudice.

CHAPTER 6

PROMISE MECHANICS

When autonomous agents make promises, they do so without any *a priori* regard for one another. Their promises do not have any notion of global design or consistency, unless they are explicitly coordinated by another agent—subordinating themselves to a planner. Thus, agents are typically not designed to form a complementary whole, and no agent necessarily has a god's eye view of every other agent's promises. As a tool for model-building, however, it is convenient to adopt such a god's eye view of the whole promise graph, imagining that the modeller is in scope of every promise.

6.1 THE LAWS OF PROMISE COMPOSITION

Let's consider how agents come together in cooperative configurations, by composing their promises into agents on a larger scale.

6.1.1 MULTIPLE PROMISES BY A SINGLE AGENT

Promise bodies are defined mathematically in terms of sets. This allows for a very general formulation, and rather simple laws of composition by set union. Moreover, any repetition in the bodies from multiple promises is only counted once, in virtue of rule 2.

Law 1 (Composition of promises). *Let b_1 and b_2 be the bodies of two promises between two agents A_1 and A_2. Then, in virtue of set semantics, and idempotence of μ-promises:*

$$A_1 \xrightarrow{b_1} A_2 \, , \, A_1 \xrightarrow{b_2} A_2 \tag{6.1}$$

$$= A_1 \xrightarrow{b_1 \, , \, b_2} A_2 \tag{6.2}$$

$$= A_1 \xrightarrow{b_1 \cup b_2} A_2 \tag{6.3}$$

6.1. THE LAWS OF PROMISE COMPOSITION

The proof of this followed trivially from the set nature of the bodies. To make this explicit, we write the promise body $\pi = \langle \tau, \chi \rangle$ and consider how these parameters combine. We can now form trivial bundles of promises between a single pair of agents by union.

$$A_1 \xrightarrow{b_1} A_2, \ A_1 \xrightarrow{b_2} A_2, \ \ldots A_1 \xrightarrow{b_N} A_2 \tag{6.4}$$

$$= A_1 \xrightarrow{b_1 \cup b_2 \cup \ldots b_N} A_2 \tag{6.5}$$

and

$$b_1 \cup b_2 = \langle \tau_1 \cup \tau_2, \chi_1 \cup \chi_2 \rangle \tag{6.6}$$

The types τ may overlap here.

6.1.2 ENSEMBLE PROMISES

We can compose *bundles* of promises by union of the promisers into an ensemble. A single agent promise to a group or ensemble, and vice versa:

$$A_1 \xrightarrow{\pi} \{A\} \tag{6.7}$$

$$\{A\} \xrightarrow{\pi} A_1 \tag{6.8}$$

A promise from every agent in one ensemble to every agent in another:

$$\{A\}_1 \xrightarrow{\pi} \{A\}_2 \tag{6.9}$$

An unspecified agent from one ensemble $\{A_1\}$ makes a promise to one agent from the ensemble $\{A_2\}$.

$$A_?(\{A\}_1) \xrightarrow{\pi} A_?(\{A_2\}). \tag{6.10}$$

6.1.3 EFFECTIVE ACTION AND COUPLING STRENGTH DURING TRANSFER OF INTENTIONAL BEHAVIOUR

Suppose one agent promises τ_1, χ_1 but the recipient of the promise accepts only τ_2, χ_2. The transferred effect of the promise binding is then given by the overlap or intersection of what is offered and what is accepted:

$$A_1 \xrightarrow{\langle \tau_1, \chi_1 \rangle} A_2, A_1 \xrightarrow{U(\langle \tau_2, \chi_2 \rangle)} A_2 \tag{6.11}$$

$$= A_1 \xrightarrow{(\langle \tau_1 \cap \tau_2, \chi_1 \cap \chi_2 \rangle = \langle \tau_1, \chi_1 \rangle)} A_2 \tag{6.12}$$

Or, put another way:

$$A_1 \xrightarrow{\langle \tau_1, \chi_1 \rangle} A_2, A_1 \xrightarrow{U(\langle \tau_2, \chi_2 \rangle)} A_2 \tag{6.13}$$

$$= A_1 \xrightarrow{(\langle \tau_1 \subseteq \tau_2, \chi_1 \subseteq \chi_2 \rangle)} A_2. \tag{6.14}$$

The semantics implied here say that a conditional promise is kept if *at least* the condition is kept. i.e. we don't need an exact match, since the condition is an enabler, not a constraint.

A binding or handshake may be described by the asymmetric notation

$$A_1 \xleftrightarrow{b} A_2 \qquad (6.15)$$

which is an irreversible replacement rule defined as follows.

$$A_1 \xrightarrow{\langle \tau_1, \chi_1 \rangle} A_2, A_2 \xrightarrow{U(\langle \tau_2, \chi_2 \rangle)} A_1 \qquad (6.16)$$

$$\equiv A_1 \xleftrightarrow{\langle \tau_1 \subseteq \tau_2, \chi_1 \subseteq \chi_2 \rangle} A_2 \qquad (6.17)$$

The asymmetric arrow denotes that the promise of something is being offered from A_1 to A_2 and there is acceptance in the opposite direction[31].

6.2 THE LAWS OF CONDITIONAL PROMISING

6.2.1 PROMISES WITH CONDITIONS ATTACHED

Not all promises are made without attached conditions. For instance, we might promise to pay for something 'cash on delivery', i.e. only after a promised item has been received. Such promises will be of central importance in discussing processes, agreements and trading. The truth or falsity of a condition C may be promised by an agent that is able to assess the condition as follows.

$$A_1 \xrightarrow{T(C)} A_2 \qquad (6.18)$$

is a promise from A_1 to A_2 of accurate information that the condition C holds. Similarly,

$$A_1 \xrightarrow{F(C)} A_2 \qquad (6.19)$$

is a promise from A_1 to A_2 of accurate information that the condition C is false.

A promise with body b made conditionally on such a Boolean condition C is now written in a notation reminiscent of conditional probabilities, as:

$$A_1 \xrightarrow{b|C} A_2. \qquad (6.20)$$

Moreover, since the condition is merely a promise itself, we can generalize this notion of conditionals to include any kind of promise that must be kept to satisfy the pre-requisite.

Definition 30 (Conditional promise). *Let b be the promise body, and X be a pre-requisite promise must be kept before b can be kept, then we denote the promise of b given X also by*

$$A_1 \xrightarrow{b|X} A_2. \qquad (6.21)$$

6.2. THE LAWS OF CONDITIONAL PROMISING

A promise that is made subject to a condition is not an assessable promise, unless the state of the condition that predicates it has also been promised, due to incomplete information. Such a promise is hence null-potent.

Law 2 (Local quenching by promised fact (exact)). *If a promise is made subject to condition C, promising simultaneously the the truth of condition leads to an unconditional promise of the full magnitude:*

$$A_1 \xrightarrow{b|C} A_2, \; A_1 \xrightarrow{\mathrm{T}(C)} A_2 \equiv A_1 \xrightarrow{b} A_2 \tag{6.22}$$

Conversely, when the condition is false, the remaining promise is rendered empty:

$$A_1 \xrightarrow{b|C} A_2, \; A_1 \xrightarrow{\mathrm{F}(C)} A_2 \equiv A_1 \xrightarrow{\emptyset} A_2 \tag{6.23}$$

Note that, since promising $T(C)$ and promising b are two different types of promise, this rule does not follow directly by the normal set algebra in section 6.1. We must deduce it from the fact that both of these promises are: (i) made to and from the same set of agents, and therefore (ii) calibrated to the same standards of trustworthiness. Thus, logically we must define it to be so.

We can generalize these results for promises of a more general nature, by the following rules.

Law 3 (Local quenching by general promise (exact)). *If a promise with body S is provided subject to the provision of a pre-requisite promise of body b, then the provision of the pre-requisite by the same agent is completely equivalent to the unconditional promise being made:*

$$A_1 \xrightarrow{S|b} A_2, \; A_1 \xrightarrow{+b} A_2 \equiv A_1 \xrightarrow{S} A_2 \tag{6.24}$$

This gives us a rewriting rule for promises made by a single agent in the promise graph. The $+$ is used to emphasize that X is being offered, and to contrast this with the next case.

Consider now the first case in which one agent assists another in keeping a promise. According to our axioms, an agent can only promise its own behaviour, not that of other agents, thus the promise basically comes from the principal agent, not the assistant, which might not even be known to the final promisee[32]. Assistance is a matter of voluntary cooperation on the part of the tertiary agent.

Law 4 (Assisted quenching of a pre-requisite promise). *If a promise with body S is provided subject to the provision of a pre-requisite promise π, then the provision of the pre-requisite by an assistant is acceptable if and only if the principal promiser also promises to acquire the service π from an assistant (promise labelled $-X$):*

$$\left. \begin{array}{c} A_T \xrightarrow{+b(\pi)} A_1, \\ \\ \end{array} \begin{array}{c} A_1 \xrightarrow{S|b(\pi)} A_2 \\ A_1 \xrightarrow{-b(\pi)} A_2 \end{array} \right\} \sim A_T \xrightarrow{+b(\pi)} A_1, A_1 \xrightarrow{S} A_2 \qquad (6.25)$$

If A_1 promises that it will deliver S conditionally on receiving $b(\pi)$, and also promises that it is receiving or will receive X from some other agent, then this is equivalent to a promise to deliver S. A_2 might assess the trustworthiness of the promise differently, given that other agents are involved, but since it is not aware of the source of X, it can only judge through the proxy of A_1. For this reason, we say that these promises are similar ($\sim$) but not identical.

We refer to this as an *assisted promise*, i.e. A_1 is assisted by third part A_T. The principal A_1 promises both that it will keep S provided π is given, and that it will both procure and use X. Then to complete the picture, a third party A_3 must provide π to A_1.

Definition 31 (Assisted promise). *As a shorthand, we shall sometimes use the notation:*

$$A_1 \xrightarrow{S|b(\pi)} A_2 \, , \, A_1 \xrightarrow{-b(\pi)} A_2 \equiv A_1 \xrightarrow{S(b)} A_2. \qquad (6.26)$$

in which the identity of the assistant is not promised.

6.2.2 COMPLEMENTARITY OF PROMISES WITH CONDITIONALS

We may extend the notion of complementarity of independent promises from section 3.11.6, with the following example.

- In the service (pull) view (a), the service provider takes ultimate responsibility by making a promise directly to the end reader, but it is a promise conditional on the behaviour of the post office whose role is to deliver the book. The positive aspect of this view is that it reflects the reality of the trading interaction. The post office is merely an assistant (see section 6.2). This is a version of causation in which the original intention is the driver for events.

- In the transport (push) view (b), we model this more closely related to the physical implementation of the promise. The service provider (bookshop) promises to pass the book to the post office, who in turn promises the reader to deliver it, assuming that it gets the book in the first place. This is a version of causation in which transactions leading to fulfilment of the promise are in focus.

6.3. MODELLING COOPERATIVE BEHAVIOUR

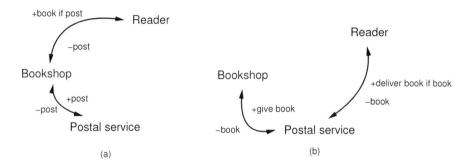

Figure 6.1: Alternative interpretations of a service interaction, in terms of a service or transport.

The first of these could be viewed a more satisfactory representation of the scenario, as it provides a deeper explanation for the events that happen to transpire, and it places the end points of the service delivery in a direct relationship with one another.

6.3 MODELLING COOPERATIVE BEHAVIOUR

To formulate a model for engineering or management, we are interested in understanding what promises would be needed to make agents deliver a desired outcome, from the perspective of some adjudicator. The incentives for promising something standard may be driven by the economics of the network in the long run[BFb]. For the time being, however, we shall assume that coordination is a property of such agents, and seek the specific patterns allowing agent nodes to promise an alignment of purpose, without violating the assumption of autonomy.

Consider the three nodes in fig. 6.2. On the surface, a promise made by node A_2

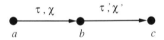

Figure 6.2: Serial composition of promise and subordination, leads to limited transitivity. The dashed arrow is implied by the $C(b)$ promise.

to node A_3 does not imply any constraint or information about a promise from node A_1 to node A_3: the nodes are autonomous. But, what if we introduce an additional promise from A_1 to A_2? Then, we can ask node A_1 to coordinate with node A_2 about its behaviour towards A_3. This requires additional promises. We shall often use a shorthand notation for a *coordination* (or subordination) promise, denoting its body $C(b)$. This is the promise to do the same as another agent with respect to a promise body b. This is not a primitive type, but we denote it by the shorthand:

$$A_1 \xrightarrow{C(b)} A_2 \qquad (6.27)$$

This means that A_2 informs A_1 about its promises of type b, and the receipt and usage of that information by A_1. This promise is a subordination for the agent because A_1 is willingly giving up its autonomy in the matter of b by agreeing to follow A_2's lead. We might call this a 'slave' promise.

Definition 32 (Promise to cooperate (subordinate)). *We define the coordination promise:*

$$A_1 \xrightarrow{+b} A_2, A_1 \xrightarrow{C(b)} A_2 \equiv \begin{cases} A_2 \xrightarrow{+\mathbf{def}(b)} A_1 \\ A_1 \xrightarrow{-\mathbf{def}(b)} A_2 \\ A_1 \xrightarrow{+b|\mathbf{def}(b)} A_2 \end{cases} \qquad (6.28)$$

$$\neq \begin{cases} A_2 \xrightarrow{+b} A_1 \\ A_1 \xrightarrow{-b} A_2 \\ A_1 \xrightarrow{+b|b} A_2 \end{cases} \qquad (6.29)$$

The slave agent promises to do the same as the master agent. This requires the master node to inform the slave of its intentions and for the slave to agree to comply with these.

Notice that, by agreeing to follow another agent's promises, this is also a voluntary subordination of the agent. It is not an obligation, because the promise can easily be withdrawn, but it is as close as we can get voluntarily.

We may now write an equivalent promise for the case in which A_1 promises A_2 that it will coordinate on the matter of b, given that A_2 promises A_3 b, and we use the symbol " , ", as above, to signify the composition of these promises.

$$\underbrace{A_1 \xrightarrow{C(b)} A_2}_{\text{'Coordinate with'}} , \underbrace{A_2 \xrightarrow{b} A_3}_{\text{Promise}} \Rightarrow A_1 \xrightarrow{b} A_3 . \qquad (6.30)$$

This allows us to deduce something about a 'global consistency' of interaction with a third-party, such as might be observed by an external observer of the agents with global knowledge. We see that this combination of promises implies that there ought logically to be a promise of type b from A_1 to A_3. It has the appearance of transitivity, but it is more subtle. In this example, we have one agent following the other. If the first agent changes

6.3. MODELLING COOPERATIVE BEHAVIOUR

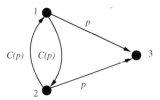

Figure 6.3: Coordination and the use of third parties to measure the equivalence of agent-nodes in a region. Agents form groups and roles by agreeing to coordinate policy.

its mind, then both promises can drift together like a 'swarm'[BF07b]. To remove this privileged leadership, we can symmetrize the promises (fig 6.3):

If two nodes pledge their promise to an external party (a form of allegiance), then they have made an implicit agreement with one another also, which binds them together. This result is important for seeing the actual groupings of roles and structures in a policy graph.

A complete structure, using the shorthand cooperative notation, looks like this:

- A_1 promises p to A_3.
- A_2 promises A_1 to collaborate about p (denote this as a promise $C(p)$).
- A_1 promises A_2 to collaborate about p (denote this as a promise $C(p)$).
- A_2 promises p to A_3

The cooperation promise will turn out to be an importance point of contact with the theory of games and voluntary cooperation[Axe97, Axe84]; this is discussed in a separate paper[BFb].

6.3.1 SUBORDINATION AND AUTONOMY

Subordination is a partial ordering on agents. Define the ordering operator

$$A <_{\text{sub}} B \tag{6.31}$$

meaning that agent A is subordinate to agent B. Then,

$$A <_{\text{sub}} B \rightarrow A \neq B \tag{6.32}$$
$$A \leq_{\text{sub}} B \leftrightarrow (A = B) \wedge (A <_{\text{sub}} B) \tag{6.33}$$
$$A \not<_{\text{sub}} A \tag{6.34}$$

And agent A is autonomous if it is the maximal element in this ranking, i.e. an arbitrary agent a satisfies $\forall a : a \leq_{\text{sub}} A$.

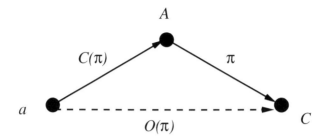

Figure 6.4: Voluntarily simulated obligation.

An agent cannot voluntarily give up its autonomy forever, since it is free to revoke such a promise, but we can easily simulate subordination voluntarily. The so-called coordination promise introduces a voluntary obligation by giving up autonomy in the matter of a promise body $b(\pi) = \langle \tau, \chi \rangle$. We already have a way of making something like an obligation in the Promise Theory. This is a limited understanding of an obligation within the promise framework. Let A be an authority and a, C be agents.

$$a \xrightarrow{C(\pi)} A , \ A \xrightarrow{\pi} C \ggcurly a \xrightarrow{O(\pi)} \blacksquare C \qquad (6.35)$$

Then by subordinating to A's wishes, a becomes obliged to promise π to C. However, $a \not<_{\text{sub}} C$.

Law 5 (Promise conflicts by cooperative/subordinate agreements). *If an agent makes a promise π of type $\tau(\pi)$, and simultaneously promises to subordinate to another agent's promise of the same type $C(\pi')$, were $\tau(\pi) = \tau(\pi')$, then there will be a conflict unless we also have $b(\pi) = b(\pi')$. In other words, if an agent subordinates to another, it cannot also make its own independent promises on the same matter that differ.*

Chapter 7

Promise Dynamics

How systems comprising agents and their promises change is the subject of promise dynamics. It comprises independently the evolution of agents and of promises.

7.1 The promise life-cycle

The promise life-cycle refers to the various states through which a promise passes, from proposal (see table 7.1). The life-cycle of a promise may now be viewed from either the perspective of the agent making the promise, or from the perspective of the promisee, or in fact any another agent that is external but in scope of the promise (see figure 7.1).

Law 6 (Consistent expiry of promises). *Once a promise is broken or otherwise enters one of its end states (invalidated, expired, etc), its life-cycle is ended, and further intentions about the subject must be described by new promises.*

From the perspective of the promisee, or other external agent in scope, we have a similar life-cycle, except that the promise is first noticed when published by the promiser (see figure 7.2).

7.2 Agent worlds and knowledge dissemination

Agents experience the world from the viewpoint of the information available to them. Each can form its own independent conclusions or ontology about the worlds of agents external to it. In order for agents to work together, it is necessary for agents to share information and mutually calibrate their expectations.

Promise State	Description
proposed	A promise statement has been written down but not yet made.
issued	A promise has been published to everyone in its scope.
noticed	A published promise is noticed by an external agent.
unknown	The promise has been published, but its outcome is unknown.
kept	The promise is assessed to have been kept.
not kept	The promise is assessed to have been not-kept.
broken	The promise is assessed to have been broken.
withdrawn	The promise is withdrawn by the promiser.
expired	The promise has passed some expiry date.
invalidated	The original assumptions allowing the promise to be kept have been invalidated by something beyond the promiser's control.
end	The time at which a promise ceases to exist.

Table 7.1: The major promise states.

Transmitted information becomes 'knowledge' when it is 'known', i.e. when it is trusted and contextualized[33]. Agents can learn and store knowledge. There is knowledge in the bodies of the promises they make. As agents repeatedly assess the promises of another, those assessments acquire the status of knowledge about the agent's promises.

7.2.1 Definition of knowledge

We must address the question of how knowledge (say of a promise) becomes known to an audience of other agents. We being by equating knowledge with information that has been assimilated by agents (see definition below) to reflect the idea that a promise is a continuously maintained state, not a mere data transaction.

Definition 33 (Agent Knowledge). *Let $K_{A_i}^\gamma$ be a set of data values, internal to agent node A_i, that have been accepted. We call $K_{A_i}^\gamma$ the knowledge of agent A_i about an observable value γ. This knowledge is not available to any other agent, unless its value is promised to them.*

We use a notation $K^\gamma, K_1^\gamma, K_2^\gamma$ as follows. Let $\mathcal{K}$ be the set of all possible variables or subjects that can be known, and consider knowledge about a specific set of values or subject $K^\gamma \in \mathcal{K}$. We use K_1^γ to mean agent A_1's knowledge of this subject γ, and K_2^γ to be A_2's knowledge about K^γ etc. Thus different agents (i) can have different knowledge about the same subject (γ). We shall use this, in particular, to discuss agents' varying type schemas.

7.2. AGENT WORLDS AND KNOWLEDGE DISSEMINATION

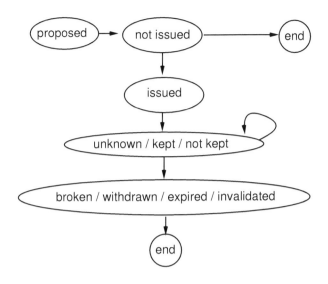

Figure 7.1: The promise life-cycle for a promiser

7.2.2 DISTRIBUTED CONSISTENCY OF KNOWLEDGE

Passing observables from agent to agent leads to a spread of information, and it can also lead to the distortion of information. How far does this loss of fidelity matter?

Law 7 (Information integrity through intermediaries). *Information transmitted by one agent represents a promise about information content. When the information is relayed through an intermediate agent, the original promise to the relay agent does not transfer to the final agent.*

This follows from the rule that agents cannot make promises on behalf of other agents. Thus the origin agent cannot promise the final agent that the intermediary will delivery the information unchanged. The intermediary might be unable or unwilling to correctly transfer information from the origin.

The scope of a promise is the simplest boundary within which there must be knowledge of a promise (see section 2.3).

Definition 34 (Consistent Knowledge). *Two agents A_s and A_r, with knowledge K_s^γ and K_r^γ, can claim to have consistent knowledge about k^γ iff they each can verify the equality of a subset of their knowledge $k_s^\gamma = k_r^\gamma$ where $k_s^\gamma \subseteq K_s^\gamma$ and $k_r^\gamma \subseteq K_r^\gamma$.*

To simplify notation henceforth, we pick a single subject and suppress the γ superscripts.

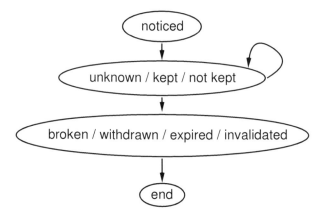

Figure 7.2: The promise life-cycle for a promisee

Theorem 2 (Consistent Knowledge). *Let A_s and A_r be any two agent nodes, and let $\pi : k$ be a promise body that implies communicating knowledge k. Nodes A_s and A_r have consistent knowledge k iff*

$$A_s \xrightarrow{+k_s} A_r \quad (7.1)$$
$$A_r \xrightarrow{-k_s} A_s \quad (7.2)$$
$$A_r \xrightarrow{+k_r} A_s \quad (7.3)$$
$$A_s \xrightarrow{-k_r} r \quad (7.4)$$
$$A_s \xrightarrow{(k_s = f(k_r, k_s))} A_r \quad (7.5)$$
$$A_r \xrightarrow{(k_r = f(k_r, k_s))} A_s \quad (7.6)$$

where $f(a, b)$ is some function of the original values that must be agreed upon.

Proof. By definition 33 the knowledge k_s of any agent S is only available to another agent node A_r iff its value is promised as in equations (7.1) and (7.3). The knowledge is not certainly received until the agents promise to accept the promised values, as in equations. (7.2) and (7.4). Now both agents know each others' values. To make their knowledge consistent, by definition 34 they must now promise each other to make their values equal. This can be done through an intermediary function $f(a, b)$ which signifies that an equilibrium is formed by some unspecified consensus. □

Consistency of knowledge is a strong concept which requires the notion of verification; at the same time it is clear that knowledge might be both available and consistent in a service oriented network, even if one cannot actually verify that consistency, e.g. if it is disseminated from a single source.

7.2. AGENT WORLDS AND KNOWLEDGE DISSEMINATION

7.2.3 ASSIMILATION OF KNOWLEDGE

Let $\mathbf{def}_s$ be the knowledge of agent A_s. Knowledge $\mathbf{def}_s$ is said to be relayed from a source A_s to a receiver agent A_r iff

$$\begin{aligned} A_s & \xrightarrow{+\mathbf{def}_s} A_r \\ A_r & \xrightarrow{-\mathbf{def}_s} A_s \end{aligned} \qquad (7.7)$$

The fact that knowledge has been passed on and received does not mean that it has been *assimilated* by the agent, i.e. that it will replace its own knowledge about $\mathbf{def}$ with this new version. For example, in peer to peer systems, such as media distribution, relay stations, routers, data are frequently relayed in this way without necessarily being retained. A new version of a song or movie could be passed on to others without replacing a local authoritative version used by the local system.

Definition 35 (Assimilated knowledge). *Knowledge $\mathbf{def}_s$ is said to be assimilated by a receiver agent A_r from a source A_s iff*

$$\begin{aligned} A_s & \xrightarrow{+\mathbf{def}_s} A_r \\ A_r & \xrightarrow{-\mathbf{def}_s} A_1 \\ A_r & \xrightarrow{(\mathbf{def}_r = \mathbf{def}_s)} A_2 \end{aligned} \qquad (7.8)$$

where A_1, A_2 could be some unspecified agents. We shall assume here that $A_1 = A_2 = A_s$ for simplicity.

An agent does not know $\mathbf{def}$ unless it is either the source of the knowledge, or it has promised to accept and assimilate the knowledge from the source. It does not matter to whom the promises to use and assimilate the data are made, as long as they are kept, though it seems natural to make the promises to A_s so that A_s can verify the transmission.

7.2.4 INTEGRITY OF INFORMATION THROUGH INTERMEDIARIES

Knowledge $\mathbf{def}$ can be passed on from agent to agent with integrity, but we must distinguish between agents that end up with a different picture of the information as a result of their local policies for receiving and relaying their knowledge.

1. Accepted from a source, ignored and passed on to a third party intact.

$$\begin{aligned} A_1 & \xrightarrow{+\mathbf{def}_1} A_2 \\ A_2 & \xrightarrow{-\mathbf{def}_1} A_1 \\ A_2 & \xrightarrow{\mathbf{def}_1} A_3 \end{aligned}$$

$$(7.9)$$

Note that the agent A_2 does not assimilate the knowledge here by making its own version $\mathbf{def}_2$ equals to $\mathbf{def}_1$, it merely passes on the value as hearsay.

2. Accepted from a source, ignored and local knowledge is then passed on to a third party instead.

$$
\begin{aligned}
A_1 &\xrightarrow{\mathbf{def}_1} A_2 \\
A_2 &\xrightarrow{U(\mathbf{def}_1)} A_1 \\
A_2 &\xrightarrow{\mathbf{def}_2} A_3
\end{aligned}
\tag{7.10}
$$

Here the agent A_2 accepts the information but instead of passing it on, passes on its own version. The source does not know that A_2 has not relayed its data with integrity.

3. Accepted and assimilated by an agent before being passed on to a third party with assurances of integrity.

$$
\begin{aligned}
A_1 &\xrightarrow{+\mathbf{def}_1} A_2 \\
A_2 &\xrightarrow{-\mathbf{def}_1} A_1 \\
A_2 &\xrightarrow{\mathbf{def}_2=\mathbf{def}_1} A_1 \\
A_2 &\xrightarrow{\mathbf{def}_2=\mathbf{def}_1} A_3 \\
A_2 &\xrightarrow{\mathbf{def}_2|\mathbf{def}_1} A_3
\end{aligned}
\tag{7.11}
$$

A_2 uses the data from A_1, assimilates it ($\mathbf{def}_2 = \mathbf{def}_1$) and promises to pass it on (conditionally $\mathbf{def}_2|\mathbf{def}_1$) if it receives $\mathbf{def}_1$. It also promises to both involved parties to assimilate the knowledge. Only in this case does the knowledge $\mathbf{def}$ become common knowledge if one continues this chain.

The first two situations are indistinguishable by the receiving agents. In the final case the promises to make $\mathbf{def}_1 = \mathbf{def}_2$ provide the information that guarantees consistency of knowledge throughout the scope of this pattern.

7.2.5 Uniformity and Common Knowledge

Given that we can define the fidelity of transmission, and the promises required to achieve it, we can now go on to define the scope of that knowledge amongst a number of agents.

The idea of common knowledge builds on an equilibration of promised information amongst a number of agents. This must take a finite time, in general, and will likely reduce the responsiveness of of any promises made by the group as a whole. Each change needs to find a new equilibrium, and continuous change may never result in complete consistency.

Definition 36 (Scope of common knowledge). *Let S be a set of source agents with consistent knowledge* **def**, *and let the set $\mathcal{A}(X, \mathbf{def})$ mean the set of nodes that have assimilated knowledge from a set of agents X. The scope $\mathcal{S}(\mathbf{def})$ of $\mathbf{def}$ is the union of S with all agents that have assimilated knowledge originating from S, i.e.*

$$\mathcal{S}(\mathbf{def}) = S \ \cup \ \mathcal{A}(S \cup \mathcal{S}(\mathbf{def}), \mathbf{def}) \tag{7.12}$$

The simplest way to achieve common knowledge is to have all agents assimilate the knowledge directly from a single (centralized) source agent. This minimizes the potential uncertainties, and the source itself can be the judge of whether the appropriate promises have been given by all agents mediating in the interaction. The scope is then simply a cooperative promise group in sense of ref. [BFa]. This is the common understanding of how a network *directory service* works. Although simplest, the centralized source model is not better than one in which data are passed on epidemically from peer to peer. The problem then is simply in knowing the boundaries of scope. Agents may thus have consistent knowledge from an authoritative source, either with or without centralization.

7.3 LAWS OF BEHAVIOUR FOR AGENTS

The nature of autonomous agents implies simple rules for how promises change agent autonomy. One expects to find laws of conservation and change in any theory of behaviour. The following may be compared with Newton's laws of motion, i.e. change of moving bodies. Notice how the flavour of the second and third laws are exchanged with respect to Newton's laws, reflecting a shift from force or obligation (push) to cooperation (pull).

Law 8 (Conservation of state). *An autonomous agent has default promises to represent the invariance or conversation of it regular state.*

1. *A continues with uniform behaviour, unless it promises to accept or subordinate to an influence from outside.*

2. *The observable behaviour of A is changed when promising to act on input from an outside source, with effective action given in section 3.11.7.*

3. *Every external influence $+b = \langle +\tau, \chi_1 \rangle$ promised by an external agent must be met by an equal and opposite promise from A $-b = \langle -\tau, \chi_2 \rangle$ in order to effect the desired change on A. If $\chi_1 \neq \chi_2$, then the interaction is of magnitude $\chi_1 \cap \chi_2$.*

7.4 DEADLOCK IN CONDITIONAL PROMISES

Promise equilibria may be viewed as pairs of bilateral promises between agents. The phenomenon is like symbiosis in biology. This mutual closure between promises is a basic topological configuration that allows for the persistence of an operational relationship. When this 'trade' of promises is stable over some time, the result is equilibrium. It is easy to show that, for agents with preferences, they correspond to Nash equilibria of two-person games[34].

Suppose now that two agents are involved in a statement of willingness to perform two independent actions A_1 and A_2 that are linked through a conditional, namely that one is predicated by the other. e.g. "I am willing to perform A_1 if I get A_2". This statement of willingness could also expressed "If I get A_2, I am willing to perform A_1". There is no causal ordering in these actions. The temporal ordering of the two agents expresses complete symmetry and there can be no causal outcome. In fact the situation is a deadlock.

In order to break this deadlock, i.e. for something to happen, or for a promise or imposition to quench one of the conditionals, we have to break this symmetry. One of the the agents must make an unconditional statement (promise) or action, i.e. it must rescind its condition and simply act in the manner of an *initial condition*. This then allows the trade of promises to be made. We shall use this in chapter 7 to explain time and deviation from equilibrium.

The occurrence of unconditional promises to act suggests that all promises to act would begin with the expression of a conditional promise, which then motivates an unconditional promise in return from the counter-agent to release the preferred course of action.

7.4. DEADLOCK IN CONDITIONAL PROMISES

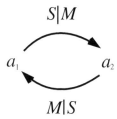

Figure 7.3: A simple exchange of actions between agents results in a causal symmetry. There is temporal equilibrium and the preferences are balanced as long as the actions are valued by one another. This is the basis of trade. One agent must concede its condition and act unconditionally to set the initial condition that generates a causal evolution.

7.4.1 FROM EQUILIBRIUM TO AN ARROW OF TIME

Consider a generator of a cycle (see fig. 7.4), as discussed in section 7.4. We can express this more carefully now, in the language of assessments.

$$a_1 \xrightarrow{\pi_S : S|A(\pi_M)} a_2 \tag{7.13}$$

$$a_2 \xrightarrow{\pi_M : M|A(\pi_S)} a_1 \tag{7.14}$$

In the previous formulation of our model for cooperation, agent a_1 promises a service S if it assesses a promise of money M to be kept. Agent a_2 promises a_1 money M if it assesses that it has received service S. This scenario represents a 'deadlock' of the kind well know in commerce. Neither party can proceed until one of them breaks the stalemate. The model is more general than just a business interaction, of course. It could

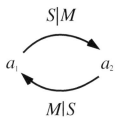

Figure 7.4: A 'cycle' generator is two deadlocked promises back to back, waiting for the deadlock symmetry to be broken.

describe any kind of pump or process.

Example 42. *Consider a car engine. There is a part that injects fuel if a flap is opened. Then there is a part that ignites it (creating motion) if the fuel is injected. These two processes are usually in deadlock, and have two possible equilibria:*

- *Nothing happens – both the fuel injector and the ignition are at rest and remain so.*

- *Both parts are in motion in a continuous cycle.*

How does the system make a transition from one equilibrium to another? Something must break the symmetry. Normally an external electric starter motor initiates motion to break this deadlock. Sometimes drivers will 'bump start' the car by pushing it and releasing the clutch if the battery is dead. In either case, there is an act of symmetry breaking.

Thus, to resolve the dilemma of how to get the pump started, the symmetry between the players must be broken by an 'initial condition'. Depending on which agent invokes this initial condition, one might assess one of two possible patterns of outcome:

$$\pi_S : A(\pi_S)A(\pi_M)A(\pi_S)A(\pi_M)\ldots \tag{7.15}$$

$$\pi_M : A(\pi_M)A(\pi_S)A(\pi_M)A(\pi_S)\ldots \tag{7.16}$$

In shorthand, we may write:

$$\pi_S : SMSMSMSM\ldots \tag{7.17}$$

$$\pi_M : MSMSMSMS\ldots \tag{7.18}$$

Selecting one of these sequences is like selecting the 'arrow of time' in the system – deciding in which direction the cycle proceeds, because reversing the order of who goes first looks very much like making the loop go clockwise or anticlockwise, depending on who goes first. In the long run, it does not matter one way or the other, as once the memory of the initial condition is long gone, the sequences are indistinguishable – they visit all the same states in equal number[35]. Boundary conditions, such as external impositions, lead to symmetry breaking[36].

Chapter 8

Reasoning About Promises

The virtue of an interaction model, based on information exchange is that we can define explicit semantics. This formalization allows one to reason about promised interactions symbolically, providing the tools for impartial analysis.

8.1 Exclusive or incompatible promises

Not all promises can be made together in a consistent way. Exclusive promises are those which cannot physically be realized at the same time. Exclusive promises cannot both be kept if the lifetimes of the promises overlap. This does not mean that incompatible promises cannot be made, it means that they cannot be resolved rationally, and could lead to problems for an agent that makes them.

> **Definition 37** (Incompatible/exclusive promises $\#$)**.** *When two promises originating from an agent are incompatible they are mutually exclusive and cannot be realized at the same time without contradiction. We write*
>
> $$A_1 \xrightarrow{b_1} A_2 \# A_1 \xrightarrow{b_2} A_3 \qquad (8.1)$$
>
> *where the $\#$ symbol means that the promises are incompatible. If $A_2 = A_3$, we may omit the agents and write $b_1 \# b_2$.*

When do we know that promises are incompatible? In general we cannot know this Each agent makes this assessment independently and arrives at its own conclusion, in its own world.

Example 43. *Promises that are incompatible:*

- *Promises for a door to be kept open and kept closed at the same time.*

- *Promises to attend meetings in Oslo and Amsterdam at the same time.*

8.2 Promise conflicts

What if several agents make identical promises? Does that mean we should expect the same from each? Each agent might intend something different from the same promise.

Example 44. *Two brands of coffee promise the smoothest cup but taste very different. Two databases, e.g. MySQL and PostgreSQL, both promise to follow the SQL standard, but are very different implementations with different focus. However, we will often perceive them to be the same 'for all intents and purposes'.*

Promises can be in conflict in a number of ways. This can affect the promiser or the promisee.

Definition 38 (Promise conflict). *Two or more promises to an agent are in conflict if at least one promise is incompatible with another. We define a conflict as the promising or exclusive promises*

$$A_1 \xrightarrow{b_1} A_2, \ A_1 \xrightarrow{b_2} A_2 \text{ with } b_1 \# b_2 \tag{8.2}$$

Clearly promising b and $\neg b$ would be excluded.

Example 45. *Think of two resistors in an electrical circuit. Resistors are marked with coloured bands which represent the promise that they make, e.g. 150 kilo-ohms, plus or minus 5 percent. Two resistors with the same markings might have difference resistances, hopefully within a ten percent range, if they keep their promises – the exact resistance of the components can only be found by using them.*

Suppose two components really do give identical promises. In that case, one is really an alias for the other. A user of the promise would not be able to distinguish the two promises from one another (see section 5.5.1). The real problem is one of imperfect information: an agent cannot truly know whether two components are identical without a 'try and see' approach.

Example 46. *Imagine a shop that promised "We sell some kind of tea". A potential user of this promise, might search for all providers of 'some kind of tea', but would these be satisfactory? Often we need to be specific and the ability to search for alternatives is not helpful.*

8.2. PROMISE CONFLICTS

From section 3.11.1, it should be clear that promises predict two kinds of inconsistent promises, or conflicts: conflicts of use − and conflicts of give + promises. The duality property of promises in section 3.11.6 further suggests that, if a conflict can occur in a + promise, then there must also be a corresponding conflict possible in a − promise. Promise conflicts relate to the existence of exclusive promises, or contention over shared resources.

The problem is one of incomplete information. A component that wants a service from another component might not have sufficient knowledge of the other agent's promises to know whether can expect its promise to be kept. This can be traced back to the way designers of the components think in terms of obligations: function, please give me 'X' – in the manner of an obligation. To design components properly, we need to think in terms of voluntary cooperation and sufficient information.

- **Conflict of give** A give-conflict is a conflict where an exclusive promise is made by an agent in a scope where several agents can see the promise. We can also think of this as a problem of uncoordinated usage.

  Suppose now two agents who want to rely on this promise (see fig. 8.1) both begin to rely on it without the consent of the giver. In our formal terms, this is an involuntary use of resources, which is an attack on the agent. The giver can promise either X or Y, but not both at the same time. Thus the fact that two users both want something at the same time causes contention for the giver.

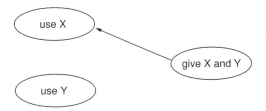

Figure 8.1: Conflict of give - promiser cannot promise both X and Y

This scenario happens in all service giving organizations, like call centers, web servers, banks, airline checking, supermarket checkouts, etc. Consider now two agents that both want access to the same promised resource X. We assume here that only one of them can use it at a time, i.e. there is only one server resource.

This kind of conflict can be avoided by the use of conditional promises. Instead of the giver promising X and Y without condition, it can promise X unless Y

or Y unless X. Which of these two is then promised depends on the order in which the users arrive. If a user agent uses X first, then Y will be unavailable, and vice versa. In computer science, this is known as transaction locking, or 'mutex' (meaning mutual exclusion). It is a way to achieve serial access to a disputed common resource.

A second workaround is to duplicate the resource itself In computer science, this is called 'threading' of the service. It is a multiple queue architecture, such as one sees at the checkout of a supermarket: rather than giving exclusive access to the cash register, one simply duplicates the service to be able to promise less contention.

- **Conflict of use**

   The dual conflict is one of inconsistent or uncoordinated giving. What if an agent relies on a promise from two or more providers in order to quench a single need? And what if the promises they keep are different? The result will be unpredictable and perhaps even superfluous; the actions of the giving agents will be uncoordinated. For instance, one could not ride two horses as the same

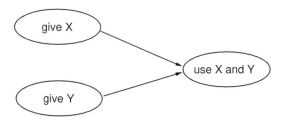

Figure 8.2: Conflict of use - cannot use both X and Y

time. The resolution here is a policy for selection from the alternatives. Such a resolution is a part of the promise made by the user, thus this kind of conflict is actually a symptom of incomplete specification. Part of a policy for handling multiple sources can be a strategy for fallback or redundancy.

We can state some simple design rules for agents that make promises to avoid exclusion, for those cases where one uses promises in a position of being able to design a system.

Rule 3 (Exclusive promises should be unique). *Exclusive promises should not be parameterizable so that they appear in multiple varieties. Each exclusive promise should*

8.2. PROMISE CONFLICTS

have a unique promise body.

Rule 4 (Advertise exclusivity). *Exclusive promises should be advertised as being exclusive, by making them conditional: e.g. I promise you exclusive access to X only if I can secure Y.*

Rule 5 (Limit scope of exclusive promises). *If one limits the scope of an exclusive promise to a single promisee, then there should be no contention for the exclusive resource.*

In designing systems, where promises are used to do bookkeeping of interactions, components making promises can often be redesigned to make promises non-exclusive. This can make systems more robust, but often at the cost of a more extensive system.

8.2.1 BREAKING PROMISES

Breaking a promise can be different to not keeping a promise. It might be an explicit contradiction. A promise broken is stronger than a promise not kept, as it implies an intended action by the agent. An intentional omission to act can also break a promise.

Definition 39 (Broken promise). *There are two essential ways in which a promise can be broken:*

1. *The agent takes an action that is incompatible with the terms of the promise.*

2. *The agent makes another promise that is incompatible with the terms of the promise.*

Note that the second case includes the case where an agent consciously or unconsciously exceeds a boundary or limitation that is expressed in the promise body, e.g. exceeds a deadline in time, or steps over a boundary in space. It is normally understood that a promise is broken by a later promise or action, but there is no need to make this restriction, and we may simply say that two promises and observations were inconsistent.

8.2.2 UNINTENDED PROMISE BREAKING

It can emerge from the logic that combinations of promises between multiple agents could lead to a promise being broken between two agents. Certain constellations of promises can lead the same agent to make and implicit or explicit promise that contradicts another that was freely given. To avoid promise conflicts one observes:

Rule 6 (Simultaneous promises must have different types). *Two parallel promises are compatible if*

$$\left.\begin{array}{c} A_1 \xrightarrow{b_1} A_2 \\ A_1 \xrightarrow{b_2} A_2 \end{array}\right\} \quad \tau(b_1) \neq \tau(b_2) \leftrightarrow \neg(b_1 \# b_2)$$

i.e. two compatible promises can be made in parallel if and only if they are not of the same type.

8.2.3 Indirection and conflicts

If the promise body contains references to agents, the logic promises is complicated considerably.

- I promise I love you.

- I promise I do not love Mary.

- I promise Mary I do not love you.

Or "I promise to you not to promise X to Y", etc.

When agents' names are absent from the promise body, the logic is very simple, following the structure of the promise graph. These levels of complexity must be built up step by step from primitive promises of the first kind. The problem cannot be avoided if we keep to the rule that the promise body may not refer to other agents, since one could reframe the promises to avoid mentioning by name and providing the data about "whom I love" as a service.

"I promise I love you, and I promise Mary I don't love you"

These are the same promise type. There is no impediment to promising this, but the result is, of course, necessarily a lie, as the conflict is detectable by the promiser.

8.3 Promise refinement and overriding

Sometimes the idea of breaking a promise with a similar promise by is too strong. The concept of inheritance of default assumptions has become commonplace in computer programming. We could conceivably allow the presence of multiple promises as long as the order in which promises override one another is defined. In fact we only need to know which of the agents has "primacy" with respect to promises of a certain type $\tau(b)$.

8.3. PROMISE REFINEMENT AND OVERRIDING

> **Definition 40** (Promise refinement). *Let b and b^* be any two promises of type $t = \tau(b) = \tau(b^*)$. We can then say that these promises of the same type*
>
> $$A_1 \xrightarrow{b^*} A_2$$
> $$A_1 \xrightarrow{b} A_2 \ldots$$
>
> *are not broken iff $b^* \subseteq b$ and promise b^* has "primacy" (denoted by the asterisk). Only one promise of type $\tau(b)$ from any promising agent can have primacy.*

In other words, a more restrictive promise does not contradict a more general promise as long as it has primacy.

Example 47. *The post office promises, by default, to deliver letters to the address printed on the front of the letter. However, this promise might be overridden by a promise to forward a letter to a change-of-address location.*

Example 48. *At a restaurant, menu item 5 might promise a dish as described in the menu specification. After indicating the he will only promise to accept item 5 if there are no carrots, menu item 5 promises to deliver the specification 5 (which includes carrots) without carrots. Thus the promise for carrots is overridden.*

Example 49. *Nearly all security access systems maintain lists called Access Control Lists that determine who can access certain resources. For instance:*

- *A supermarket promises to sell goods to customers, but only promises to sell beer to customers over 18 years old.*

- *An airport departure terminal promises to admit all persons with boarding cards. The airport lounge (which is part of the departure terminal) promises to admit only passengers with business class tickets or a gold card.*

- *A museum promises to admit anyone with a valid ticket, but the museum shop (which is part of the museum) promises to admit anyone.*

In many of these cases the issues are indicative of a deeper level of granularity in the promising agent. If we take the example of the museum, there is no need to override a promise if we recognize that the museum 'agent' can be decomposed into two different agents, one for the shop and one for the rest. Then each makes clear promises with no need to override. In the case of the supermarket, it could be considered a modelling error to view the supermarket as the promiser, with so much internal structure. A better approach might be to make each type of good promise availability to different customers, then there is no overlap or need to override. The main cause of overriding is a desire to avoid modelling internal details; however, then one questions the motivation.

This concept of overriding promises is somewhat undesirable, since it only complicates the logic of consistency. We believe it should only be used in exceptional circumstances when designing policies.

8.4 AGREEMENT

Agreement implies a shared belief, i.e. the acceptance of a proposal or hypothesis by several agents. We may define agreement amongst agents as a promise to accept a shared promise proposal.

8.4.1 AGREEMENT AS PROMISES

If agents promise to use or accept a proposal, then they attest to their belief in it.

Definition 41 (Agreement). *A promise pattern in which one or more agents $\{A_n\}$ use or accept a proposal X that is in their mutual scope. They agree iff:*

$$\{A_n\} \xrightarrow{U(X)} A_{\text{observer}}. \tag{8.3}$$

Agents within the ensemble $\{A_n\}$ may or may not know about their common state of agreement. However the observer observes a common role by association (see section 4.4.1) hence calibrating their agreement.

Example 50. *For example, two parties can agree that $2 + 2 = 4$, and this is assessed by a third party observer who calibrates their belief.*

$$A_1 \xrightarrow{U(2+2=4)} A_{\text{observer}} \tag{8.4}$$

$$A_2 \xrightarrow{U(2+2=4)} A_{\text{observer}} \tag{8.5}$$

If the proposal itself originated from another agent, then one could only deduce that the agents were also in agreement with that agent if the originator had also promised to accept the proposal.

8.4.2 CONTRACTUAL AGREEMENT

The interpretation of agreement above solves one of the long-running controversies in philosophy about whether contracts are bilateral sets of promises[37]. We may now see, rather simply, that a contract is a set of bilateral *promise proposals* (not yet promises). These come into force at the moment that all parties promise to accept the proposal[38].

8.4. AGREEMENT

Definition 42 (Contract). *A bundle of promise proposals between a number of agents, that is intended to serve as the body of an agreement.*

A contract is not an agreement until it has been accepted by all parties, i.e. until the parties agree to it[39].

Example 51. *Service (Level) Agreements are legal documents describing a relationship between a provider and a consumer. The body of the agreement consists of the promised behaviours surrounding the delivery and consumption of a service. From this we may define a Service Level Agreement as an agreement between two agents whose body describes a contract for service delivery and consumption.*

8.4.3 CONTRACTS AND SIGNING

The act of signing a proposal is very similar to agreement. It is a realization of an act that has the effect of making a promise.

Definition 43 (Signing a proposal). *A promise by an agent to use or accept a proposed bundle of promises. It is realized typically by attaching an agent's unique identity e.g. personal mark, DNA, digital key.*

Example 52. *When a client signs an insurance contract, he or she agrees about the proposed content, and the contract comes into force by signing. What about the counter-signing by the insurance company? Sometimes this is explicit, sometimes implicit. If there is no explicity counter-signing, one presumes that the act of writing the contract proposal, on company letterhead counts as a sufficient signature or promise to accept the agreement*[40].

8.4.4 COOPERATIVE AGREEMENT OR TREATY

If two agents agree to agree to behave according to the same standard as one another, then their common agreement is the intersection.

Definition 44 (Treaty Agreement). *A mutual agreement to behave in the same way.*

$$a \xrightarrow{C(\pi_1)} b, b \xrightarrow{C(\pi_2)} a \Longrightarrow a \xleftrightarrow{\pi_1 \cap \pi_2} b \tag{8.6}$$

This clarifies the process and limitations of an agreement. Each party might promise asymmetrically to keep certain behaviours, but the overlap is the only part on which they agree. Note that neither party can be subordinate to the other in such an agreement.

Chapter 9

The Value of Promises

Each agent is free to place its own value on the giving or receiving of a promise. This valuation may be quantitative or qualitative, numerical or heuristic. A promise valuation is a special form of assessment.

9.1 Value function of a promise

How much is a promise worth to each of the stakeholders in a promise?

- To the promiser?
- To the promisee?
- To others in scope?

Being worth something usually implies that knowledge of the assessment enables something that was not possible without it.

Example 53. *The cost of a promise is a negative valuation incurred by the making of a promise. This might be a cost in the agent's time, in space or other resource, in money, or in exchange value. A promiser might sometimes confuse the cost of the promise with the value it wants to recoup by reciprocation.*

Example 54. *A promises $+X$ to A'. A assesses that the value of the promise lies in good relations with A' (a small price to pay). A' assesses that a need it had was fulfilled. An external observer assesses that its safety has increased due to the good relations between A and A', which could mean lower cost in security expenditure.*

Example 55. *A special deal promises a free voucher worth 100 dollars to be used for the over-priced entry fee to Fantasy World amusement park. The voucher reduces the price to a normal level compared to other similar parks. What is the value of the voucher promise?*

Example 56. *Most authors will tell you that the value of writing a book is not measured in riches or royalties, but in the doors it opens to other opportunities. A book plays the role of targeted advertising, leading to invitations to speak or consult by the author.*

9.2 VALUATION FUNCTION

We may assume that each agent A_i possesses a *valuation function* $v_i(\cdot)$ which it can apply to promises given and received in order to determine their value. An agent A_i can only evaluate this function on a promise that it knows about, i.e. one that

- It makes $v_i(A_i \xrightarrow{\pi} A_j)$,
- That it receives $v_i(A_j \xrightarrow{\pi} A_i)$, or
- One hat is in its scope $v_i(A_j \xrightarrow[A_i]{\pi} A_k)$.

The currency of the value v_i is private to the world of agent A_i concerned. A promise need not be given or used by any agent in order to be considered valuable by an agent[41].

Example 57. *Decorations, peacock feathers, and attractive packaging do not affect the ability of an agent to fulfill its purpose in an immediate functional model (though they do affect the promises of the species at a systemic level). However, they might be valued attributes anyway, e.g. for marketing. Deeper models might turn aesthetic qualities into functional ones at a system level, but this should not be necessary to capture of the economics of promises.*

Law 9 (Promise value is an arbitrary assignment of an agent). *Agents determine the value of promises based in internal assessments. Values can be suggested by other agents by imposition or promise, but these have to be accepted and calibrated to an agent's own expectations.*

Example 58. *The value of a currency is exactly what someone is willing to give you for it. A dollar may be worth a loaf of bread one day, or a car the next. Nothing has intrinsic value. Different kinds of money can be compared arbitrarily. Sometimes money is useless: if you don't have enough to pay for your whole college fee, you can't use only half of it.*

Example 59. *Sometimes value is quantized: all or nothing. Physics standardizes value in units of energy. Different kinds of energy can only be compared arbitrarily. Sometimes energy is useless—e.g. if you don't have enough to escape into orbit then it's of no value.*

9.3 Mean Time to Keep a Promise

Often we need to characterize an agent's assessment of how long it took for a promise to be kept. This can vary from agent to agent, due to delays in propagation of action and observation. The notation:

> **Definition 45** (Time to Keep a Promise). *The estimated timescale or sometime 'Mean Time To Keep A Promise' is an assessment which may be based on*
>
> $$T_i(\pi) = \alpha_i^T(\pi) \tag{9.1}$$
>
> *Sometimes the notation $\Delta T_i(\pi)$ is used to emphasize a relative time interval rather than a timescale.*

Example 60 (MTBF and MTTR). *In fault diagnosis, one often speaks of the two-state model in which components promise either to be in working order or broken, i.e. in a state of failure or repair, where*

$$MTTR + MTBF = \Delta T_{total}, \tag{9.2}$$

where MTBF is the 'Mean Time Before Failure', when the system keeps its promise, and MTTR is the 'Mean Time to Repair' during which the system fails to keep its promise.

$$MTBF = \Delta(\pi) \tag{9.3}$$
$$MTTR = \Delta(\neg\pi) = \Delta T_{total} - MTBF. \tag{9.4}$$

These are learned averages, as one can never promise precisely how long a repair or a failure will take to emerge.

9.4 Some More Possible Valuations

As examples we list some suggested valuations that agents make.

1. The benefit of having the promise (to the promisee).

2. The benefit if the promise is kept once or multiple times.

9.4. SOME MORE POSSIBLE VALUATIONS

3. The benefit of making the promise (to the promiser), e.g. expected return on investment.

4. The benefit of having a promise kept between two other parties (to someone in scope), e.g. a peace treaty.

5. A ranking of promise importance, e.g. policy decision.

6. Expected time to keep or deliver on the promise (time to value).

7. Quality of promise outcome.

8. Cost of keeping or using the promise.

9. The assessment of one's dependence on the promise, i.e. vulnerability.

10. Reliability, trust, confidence of keeping the promise, promiser or promisee.

11. Social benefits from interacting with a community of agents, e.g. safety, security.

12. Risk of keeping or not keeping a promise, e.g. a threat.

13. Trust that a promise will lead to a desired outcome.

Chapter 10

Trust and Promise Keeping

Trust and promises are inextricably entwined. The amount of trust we assign an agent is a valuation, i.e. a policy decision, but it may grow and decline with experience of their behaviour. We can define trust from the perspective of autonomy, by relating it directly on the concept of how reliably a promise is kept. Trust and assessments are therefore closely related.

10.1 Trust and autonomous promises

Humans trust both other humans and inanimate objects ("I trust my computer to give the right answer")[42]. More precisely, we trust promises—including facts, assertions, and other claims about past and future, and more. Trust is a valuation, made autonomously, by an individual agent about another agent or its promise. An agent that is known to keep its promises might be considered trustworthy. It seems natural then to associate a measure of trust with an agent's expectation of the expectation of another agent in keeping its promises. The usefulness of promises here is that they encapsulate the intent that underlies trust.

Proposal 1 (Trust). *An* agent's expectation *that a promise will be kept. It may be assigned a value lying between 0 and 1, in the manner of a Bayesian probability.*

This proposal has a number of positive qualities. To begin with it separates the *experiential* aspect of trust from the *nature of the actions* on which it is based. Thus in terms of philosophy of science, it makes a clean distinction between empirical knowledge (expectation) and theoretical knowledge (a promise)[43].

10.2 A COMPATIBLE NOTATION FOR TRUST

We write trust relationships similarly to promises:

$$S[T] \; \mathbf{Trusts} \overset{b}{\;} R[U] \tag{10.1}$$

meaning that S trusts R to ensure that T keeps a promise of b to U.

In most cases, this is too much generality. In a world of autonomous agents, no agent would expect agent S to be able to ensure anything about agent T's behaviour. The more common case is therefore with only three parties

$$A_1[A_2] \; \mathbf{Trusts} \overset{b}{\;} A_2[A_3] \tag{10.2}$$

i.e. agent A_1 trusts agent A_2 to keep its promise towards some third-party agent A_3. Indeed, in most cases A_3 might also be identified with A_1:

$$A_1[A_2] \; \mathbf{Trusts} \overset{b}{\;} A_2[A_1] \tag{10.3}$$

which, in turn, can be simplified to

$$A_1 \; \mathbf{Trusts} \overset{b}{\;} A_2. \tag{10.4}$$

In this case, trust is seen to be a dual concept to that of a promise. If we use the notation of ref. [BFb], then we can write trust as one possible valuation $v : \pi \to [0, 1]$ by A_1 of the promise made by A_2 to it:

$$A_1[A_2] \; \mathbf{Trusts} \overset{b}{\;} A_2[A_1] \leftrightarrow v_1(A_2 \overset{b}{\to} A_1) \tag{10.5}$$

This is then a valuation on a par with economic valuations of how much a promise is worth to an agent[BFb]. The recipient of a promise can only make such a valuation if it knows that the promise has been made.

Proposal 2. *Trust of an agent S by another agent R can exist if agent R is informed that agent S has made a promise to it in the past, or if the recipient of the promise R is able to infer by indirect means that S has made such a promise.*

A natural valuation is an agent's estimate of the expectation value that the promise will be kept, i.e. an estimate of the reliability of the agent's promise.

$$A_1[A_2] \; \mathbf{Trusts} \overset{b}{\;} A_2[A_1] \overset{P}{\equiv} E_1(A_2 \overset{b}{\to} A_1) \tag{10.6}$$

where $\overset{P}{\equiv}$ means 'is defined by policy as', and the expectation value $E_R(\cdot)$, for agent R is defined by the agent A_1.

Some examples are given below, for further details see [BB06].

Example 61. *I trust my computer to give me the right answer.*

$$\text{Me} \stackrel{\text{right answer}}{\textbf{Trusts}} \text{Computer} \stackrel{P}{\equiv} E_{\text{Me}}(\text{Computer} \xrightarrow{\text{answer}} \text{Me}) \qquad (10.7)$$

i.e. I expect that the computer will keep its (implicit) promise to furnish me with the correct answer. Alternatively:

$$[\text{Me}][\text{Computer}] \stackrel{\text{answer}}{\textbf{Trusts}} [\text{Vendor}][\text{Me}]$$
$$\stackrel{P}{\equiv} E_{\text{Me}}\left([\text{Vendor}][\text{Computer}] \xrightarrow{\text{Answer}} [\text{Me}][\text{Me}]\right) \qquad (10.8)$$

i.e. I expect the promise by the vendor to me, to make the computer give me the right answer, will be kept.

Example 62. *I trust the identity of a person (e.g. by presence, public key or signature).*

$$\text{Me} \stackrel{\text{Authentic}}{\textbf{Trusts}} \text{Signature} \stackrel{P}{\equiv} E_{\text{Me}}(\text{Signature} \xrightarrow{\text{Authentic}} \text{Me}) \qquad (10.9)$$

Or:

$$\text{Me}[\text{Signature}] \stackrel{\text{Authentic}}{\textbf{Trusts}} \text{Certifier}[\text{Me}]$$
$$\stackrel{P}{\equiv} E_{\text{Me}}(\text{Certifier}[\text{Signature}] \xrightarrow{\text{Authentic}} \text{Me}) \qquad (10.10)$$

i.e. I trust a Certifying Agency to ensure that the implicit promise made by the credential to represent someone is kept.

A third interpretation is that the trust of the key is based on the promise to verify its authenticity, on demand. This is the common understanding of the "trusted third party".

$$\text{Me} \stackrel{\text{verify key}}{\textbf{Trusts}} \text{Certifier} \stackrel{P}{\equiv} E_{\text{Me}}\left(\text{Certifier} \xrightarrow{\text{verify key}} \text{Me}\right) \qquad (10.11)$$

i.e. I trust that the key has been authorized and is verifiable by the named Certification Agency. This last case avoids the problem of why one should trust the Certifying Agency, since it refers only to the verification service itself.

Example 63. *I trust money notes to be legal tender:*

$$\text{Me} \stackrel{\text{legal}}{\textbf{Trusts}} \text{Note} \stackrel{P}{\equiv} E_{\text{Me}}\left(\text{Cashier} \xrightarrow{\text{gold} \vee \text{note}} \text{Me}\right) \qquad (10.12)$$

we expect that the chief cashier will remunerate us in gold on presenting the note. Alternatively, we assume that others will promise to accept the note as money in the United Kingdom (UK):

$$\text{Me} \stackrel{\text{legal}}{\textbf{Trusts}} \text{Note} \stackrel{P}{\equiv} E_{\text{Me}}\left(S \xrightarrow{U(\text{note})} \text{Me}\right), \quad \forall S \in \text{UK} \qquad (10.13)$$

10.3. HOW PROMISES GUIDE EXPECTATION

Interestingly neither dollars nor Euros make any much promise. Rather, the dollar bill merely claims "In God we trust"[44].

Example 64. *Trust in family and friends.*

For some bundle of one or more promises $\mathcal{P}^*$ (see notation $\Rightarrow$ in section 4.1),

$$\text{Me } \mathbf{Trusts} \stackrel{\mathcal{P}^*}{\longrightarrow} \{\text{Family}\} \stackrel{P}{\equiv} E_{\text{Me}}\left(\{\text{Family}\} \stackrel{\mathcal{P}^*}{\Longrightarrow} A_i\right) \quad (10.14)$$

i.e. for some arbitrary set of promises, we form an expectation about the likelihood that family and friends would keep their respective promises to the respective promisees. These promises might, in fact, be hypothetical and the evaluations mere beliefs. On the other hand, we might possess actual knowledge of these transactions, and base judgement on the word of one of these family/friend members to keep their promises to the third parties:

$$\text{Me } \mathbf{Trusts} \stackrel{\mathcal{P}^*}{\longrightarrow} \{\text{Family}\} \stackrel{P}{\equiv} E_{\text{Me}}\left(\{\text{Family}\} \stackrel{\mathcal{P}^*}{\Longrightarrow} \text{Me}[A_i]\right) \quad (10.15)$$

Example 65. *A trustworthy employee.*

$$\text{Boss } \mathbf{Trusts} \stackrel{\text{Deliver}}{\longrightarrow} \text{Employee} \stackrel{P}{\equiv} E_{\text{Boss}}(\text{Employee} \stackrel{\text{Deliver}}{\longrightarrow} \text{Boss}) \quad (10.16)$$

Example 66. *I trust you will behave better from now on!*

$$\text{Me } \mathbf{Trusts} \stackrel{\text{Do better}}{\longrightarrow} \text{You} \stackrel{P}{\equiv} E_{\text{Me}}\left(\text{You} \stackrel{\text{Do better}}{\longrightarrow} \text{Me}\right) \quad (10.17)$$

10.3 How Promises Guide Expectation

What kind of policy should be employed in defining the expectation of future behaviour? Probability Theory is built on the assumption that past evidence can motivate a prediction of the future. At the heart of this is an assumption that the world is basically constant. However, future prediction is the essence of gambling: there are scenarios in which evidence of the past is not an adequate guide to future behaviour. An agent might also look elsewhere for guidance.

- *Initialization*: An agent of which we have initially no experience might be assigned an initial trust value of 1, $\frac{1}{2}$, or 0 if we are respectively trusting, neutral or untrusting by nature.

- *Experience*: One's own direct experience of a service or promise has primacy as a basis for trusting an agent in a network. However, an optimistic agent might choose not to allow the past to rule the future, believing that agents can change their behaviour, e.g. "the agent was having a bad day".

- *Advice*: An agent might feel that it is not the best judge and seek the advice of a reputable or trustworthy agent. "Let's see what X thinks".

- *Reputation*: Someone else's experience with a promise can serve as an initial value for our own trust.

- *Broken trust—'Damnation'*: Some agents believe that, if an agent fails even once to fulfill a promise, then it is completely un-trustworthy. This extreme policy seems excessive, since there might be reasons beyond the control of the agent that prevent it from delivering on its promise.

If we lack any evidence at all about the trustworthiness of an agent with respect to a given promise, we might adopt a policy of using the agent's record of keeping other kinds of promises. In the absence of direct evidence of type $t(b)$, in a promise body b, an agent may use a default policy, e.g. combining values from other types as an initial estimate.

Part II

Applications

CHAPTER 11

WORKFLOWS AND END-TO-END DELIVERY

Promises allow us to examine simple models of cooperation and compare them. The logistic chain is ubiquitous in business and organizational logistics. It represents a continuous delivery, workflow, or production pipeline, with widespread importance to many organizations that deliver goods and services.

11.1 INTERMEDIARIES OR PROXIES

Suppose the giver of a promise requires the assistance of other intermediary agents in order to keep its promise. The basic algebra of promise conditionals may be used to examine how such an agent can promise delivery of goods and services through a number of intermediaries[45]. This is the so-called *end-to-end delivery problem*[Bur08, Gao09] and it forms the basis of many solutions. Schematically, we typically imagine the problem as a story line narrative, as in figure 11.1.

Example 67. *Intermediary agents include postal delivery agents, transportation agents, publishers, cabling infrastructure, waiters, shops, service agents, and actors, to name a few.*

Figure 11.1 is a common way of representing processes as a workflow. In practice, it's a simplistic after-the-fact idealization of what actually happens during a single enactment of events to keep promises. It may not be a good representation of process continuity. What happens if the good or service is not delivered? How does this distinguish discrete

Figure 11.1: Schematic service delivery through a proxy or intermediate agent.

packages (like books) from continuous streams (like video)? The simple story line in the figure is not helpful from an operational perspective.

11.2 SIMPLE PROCESS MODELS IN HUMAN SYSTEMS

To see how autonomous agent promise modelling is different from the more commonly used orchestrated action modelling, it's helpful look at a couple of mundane examples in human systems. What makes human systems different is the amount of interpretation and internal reasoning the agents engage in.

11.2.1 BUYING MILK

Consider the interaction between a customer and a store. Suppose the customer wants to go to a shop to buy milk. We might start out by thinking simply that a shop promises to sell milk and the customer promises to buy it, so we might have:

$$\text{Shop} \xrightarrow{+\text{milk}} \text{Customer} \qquad (11.1)$$
$$\text{Customer} \xrightarrow{-\text{milk}} \text{Shop}. \qquad (11.2)$$

This expresses the directed intent, and it complies with the autonomy of the agents, but it isn't how shopping really works. We could add the dependency on money, and some more details, but we need to step back and think more carefully about how shops build a hierarchy of promises to invite customers to buy products, and the role of promises and impositions.

A shop typically begins its level of invitation with opening hours:

$$\text{Shop} \xrightarrow{+\text{open 7-11}} \text{Customer} \qquad (11.3)$$
$$\text{Customer} \xrightarrow{-\text{open 10:30}} \text{Shop}. \qquad (11.4)$$

The overlap opens a channel for general interaction between the shop and the customer. The general promise is designed like a fishing net to cover most of the times when customers might accept its availability, based on some general societal conventions. The shop also implicitly promises (by virtue of convention) that it accepts money, at least in some forms of currency:

$$\text{Shop} \xrightarrow{-\text{pay cash, card}} \text{Customer} \qquad (11.5)$$

11.2. SIMPLE PROCESS MODELS IN HUMAN SYSTEMS

The customer will pay conditionally on the promise of desired goods being kept:

$$\text{Customer} \xrightarrow{+\text{pay with card} \mid \text{goods}} \text{Shop}. \tag{11.6}$$

The shop will offer the goods assuming the customer will pay:

$$\text{Shop} \xrightarrow{+\text{goods} \mid \text{pay with} *} \text{Customer} \tag{11.7}$$

There is now a deadlock in the promises: both shop and customer stand willing to exchange, but no one is promising to go first (unconditionally). To break this deadlock symmetry, the customer will select and pay conditionally, having been invited into a scenario that is trusted. So the customer imposes on the shop for milk:

$$\text{Customer} \xrightarrow{-\text{milk}} \blacksquare \text{ Shop}. \tag{11.8}$$

This is an imposition, because the choice was not previously announced or ordered; but it's an imposition that is accepted by the shop, which has basically promised to plan for it. We see how the promises become narrower as invitation turns into outcome.

11.2.2 BOOKING A DOCTOR APPOINTMENT

Getting help from a doctor is a similar hierarchy of promises that proceed by layers on narrowing invitation. A doctor's practice starts by announcing free appointment slots in its calendar:

$$\text{Doctor} \xrightarrow{+\text{appointments}} \text{Patient} \tag{11.9}$$

$$\text{Patient} \xrightarrow{-\text{appointment}} \text{Doctor}. \tag{11.10}$$

The patient accepts on of these slots. The doctor promises treatment, conditionally on something being wrong, and may promise advice unconditionally:

$$\text{Doctor} \xrightarrow{+\text{treatment} \mid \text{condition}} \text{Patient} \tag{11.11}$$

$$\text{Doctor} \xrightarrow{+\text{advice}} \text{Patient} \tag{11.12}$$

The patient may promise a condition, to be assessed by the doctor.

$$\text{Patient} \xrightarrow{+\text{condition}} \text{Doctor}. \tag{11.13}$$

The doctor may be willing or able to accept a set of conditions

$$\text{Doctor} \xrightarrow{-\text{conditions}} \text{Patient}. \tag{11.14}$$

So, if the patient condition overlaps with the doctor's set of accepted conditions, treatment will be promised. Finally, to complete the bindings, the patient would promise to accept the treatment:

$$\text{Patient} \xrightarrow{-\text{treatment}} \text{Doctor}. \tag{11.15}$$

11.2.3 Hiring a Contractor or Employee

On the face of it, hiring someone for a job or project seems like a straightforward task. People promise certain skills, on their CV or media profile, and companies promise what they need. Sometimes hiring agencies act as go-betweens but the essence is to look for a binding between promised skills and required skills.

The problem with this is that companies don't really know what they need, and being able to promise a narrow skill associated with a project does not make an employee into a useful resource. Hiring is an uncertain process, and the skills needed for a job are not always correctly promised. They can also change as time goes by. Recruiting from within one's own company to a particular project is less risky, because one already knows the person's broader palette of promisable attributes. The interaction between the project manager and the potential worker plays a role.

- The classic top-down approach would be for a manager to assign tasks T_i onto workers A_i:

$$\text{Manager} \xrightarrow{+T_i} \blacksquare\, A_i. \qquad (11.16)$$

Without knowing whether agents were able and willing, this has only an uncertain outcome. By using imposition, the manager ignores possible dependencies behind the scenes. What conditions might actually be necessary to complete the tasks?

- The alternative is for a manager to invite agents to make an offer of what they can promise (this is how contracts are often awarded). The agents then promise what they can do:

$$A_i \xrightarrow{+X_i} \text{Manager}. \qquad (11.17)$$

The manager then promises to accept and marshal these offers to fit a solution. This means that the combination of the manager and the accepted contractors can form a superagent to make a more realistic conditional promise to the company about the project:

$$\{\text{Manager}, A_i\} \xrightarrow{+\text{project} \mid X_i} \text{company} \qquad (11.18)$$

$$\{\text{Manager}, A_i\} \xrightarrow{+X_i} \text{company}. \qquad (11.19)$$

This reveals all of the skills available, the fact that they have been acquired, and implies that a solution for combining them has been found.

11.2.4 MILITARY SYSTEMS

Military forces operate as far as possible like machines. Soldiers voluntarily give up their autonomy in order to imitate a machine-like operation. The idea is that this leads to a quasi-determinism, as far as possible free of the human part of uncertainty.

11.3 PROMISING STATE VERSUS PROMISING CHANGE

Consider these two different formulations of a promise to ensure that an agent receives something we call X.

- Promise to give an agent X (push action).
- Promise that the agent shall have X (achieve state).

The first of these represents a discrete one-time delivery of X to the agent, which is then satisfied and is never repeated. Since the promise is idempotent, to give another X, or a replacement X would require a different promise altogether, in accordance with section 3.13.

The second promise represents a continuous attempt to ensure that the agent always has X. The statement can persist for indefinite time. Since it maintains a state, idempotence of the promise merely ensures that the total number of X should be one at all times.

The first of these alternatives formulates its goal as a relative change. The latter describes an absolute state specification. We see that continuity of state is associated with absolute promises.

Rule 7 (Absolute promises versus relative impositions). *Use relative state impositions to describe singular changes. Use absolute state promises for continuously assessed promises about lasting, steady state equilibria.*

Although we envisage agents in a continual state of readiness to keep promises, push promises may be quenched only once, since the states they depend are lost in a timeline. Continuous promises kept on demand last for extended times. We may use these extreme positions to model multi-agent workflows that are either one-off triggered events or continuous flows.

11.4 DELIVERY CHAINS

We begin with a single intermediary agent or proxy, as in figure 11.1, in which the agent forwards a promised good or service without alteration, i.e. a relay.

11.4.1 RELATIVE DELTA-PUSH DELIVERY PATTERN

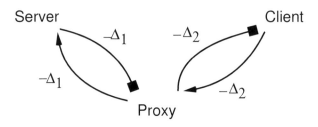

Figure 11.2: Imposition based workflow, or 'fire and forget'.

In an event-driven view (see figure 11.2), the server tries to push a change or delivery by imposing it on the delivery agent. This is sometimes called 'fire and forget', as the trigger event does not seek any assurance of final outcome. The intermediary or proxy promises to accept such impositions, using (3.40).

$$\text{Server} \xrightarrow{-\Delta_1} \blacksquare \text{Proxy} \qquad (11.20)$$
$$\text{Proxy} \xrightarrow{-\Delta_1} \text{Server} \qquad (11.21)$$
$$\text{Proxy} \xrightarrow{-\Delta_2} \blacksquare \text{Client} \qquad (11.22)$$
$$\text{Client} \xrightarrow{-\Delta_2} \text{Proxy}. \qquad (11.23)$$

1. In (11.20), the service source pushes out imposition Δ_1, meaning 'Take this package and deliver it to Client'.

2. The proxy agent promises to accept (11.21), either before or in response to the imposition, then it imposes receipt of the package on the client (11.22) where Δ_2 is 'take this package'.

3. If the Client promises to take the package (11.23), then we are done.

This view is in based on ephemeral impositions. These are fragile. If the chain fails somewhere, it would require a new imposition to restart the intended sequence of events. This is a weakness in the relative model, since it is driven by a particular initial state (retarded boundary condition). The server ends up with no assurances about what happens at the client end.

11.4.2 DELIVERY PATTERN FOR ONE INTERMEDIARY

In a desired end-state view we may formulate the promises shown in figure 11.3. The

11.4. DELIVERY CHAINS

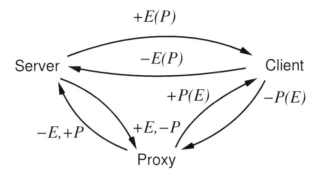

Figure 11.3: Desired state, self-maintaining workflow, or assured delivery.

server promises the client end-state E: 'You will have package' and this is conditional on 'if the delivery agent promises π_P'. The Server also promises to use a promise to delivery the package from a proxy, where P is 'Deliver package to Client'. Recall that the combination of these two promises may be written more simply, by the law of conditional assistance in section 6.2, as:

$$\left. \begin{array}{c} \pi_E : \text{Server} \xrightarrow{+E|P} \text{Client} \\ \pi_P : \text{Server} \xrightarrow{-P} \text{Client} \end{array} \right\} \equiv \text{Server} \xrightarrow{+E(P)} \text{Client}. \qquad (11.24)$$

i.e. a shorthand for 'I will deliver the end-state if some proxy helps me $+E|P$, and I promise you that I am accepting such help $-P$', which looks like a function $E(P)$, for 'end-state depending on my proxy'. This promise from the server to the client represents a virtual interface between the two, which could not be represented at all in the push-imposition model. It represents is the effective promise from the server to the client, and the client accepts. The full collaboration now takes the form:

$$\text{Server} \xrightarrow{+E(P)} \text{Client} \qquad (11.25)$$
$$\text{Server} \xrightarrow{-P,+E} \text{Proxy} \qquad (11.26)$$
$$\text{Proxy} \xrightarrow{+P,-E} \text{Server} \qquad (11.27)$$
$$\text{Proxy} \xrightarrow{+P(E)} \text{Client} \qquad (11.28)$$
$$\text{Client} \xrightarrow{-E(P)} \text{Server} \qquad (11.29)$$
$$\text{Client} \xrightarrow{-P(E)} \text{Proxy} \qquad (11.30)$$

Notice the symmetries between $\pm$ in the promise collaboration of equilibrium state, and between E, P indicating the complementarity of the promises. The Server promises its client 'I will give you E if the delivery agent promises you P'. The delivery agent says

'I will deliver P if I receive E from the Server'. Both agencies are promising the client something that requires them to work together, and the only different between them from the client's viewpoint is the realization of how the promises are kept.

1. In (11.25) and (11.29) the Server promises to deliver the end-state E via the proxy delivery agent.

2. To accomplish this, the Server promises to hand over $+E$ to the proxy, and the proxy promises to accept such transactions $-E$ in (11.26).

3. The Proxy promises the Server that is can deliver $+P$ to the client in (11.27).

4. In (11.28), the delivery agent promises to deliver what it received from the Server $+P(E)$, because it needs confirmation $-P(E)$ from the client in (11.30) that it is okay to deliver, assuming that the promised end state π_E was kept, or equivalently that it will deliver when the client makes its pull request to acquire the state.

With this constellation of promises, the continuous delivery and maintenance of E across a single intermediary agent is assured. The proof of continuity may be seen by noting that no promise can terminate, because its work is never done. Moreover, if any promise is not kept, the continuation of the promise assures that best effort will maintain the repair of the situation.

We can summarize this behaviour by a thumb rule.

Rule 8 (Promise Continuity). *When a system promises continuous operation, none of the promises may become invalid or be removed from the picture as a result of a failure to keep any promise, i.e. promises are described for all times and conditions[46], in a continuous steady state.*

What makes this second promise-only view important is that it allows us to define the virtual interface between client and server, or customer and provider—and to abstract away the helper agent. The picture in figure 11.1 is not the picture that client and server imagine. Rather the preferred picture is that of figure 11.4 Here the intermediate agent is treated as a hidden detail (like a computer subroutine), concealed within the operation of the system.

11.4.3 DESIRED END-STATE DELIVERY PATTERN FOR MULTIPLE INTERMEDIARIES

We may generalize the case of the intermediate hand-over agent into an N intermediary logistic chain. This is a classic case of transport routing, where the intermediate agents

11.4. DELIVERY CHAINS

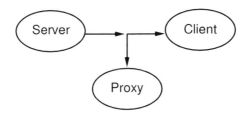

Figure 11.4: Schematic service delivery through a hidden proxy.

don't more than pass along a good or service S from a server to a client, through a chain of proxies.

To avoid too many lines in the graph, we abbreviate arrows for giving and receiving promises using the $\pm$ notation implicitly, where the forward arrow is the $+$ sign and the $-$ promise implies an arrow back in the opposite direction to accept the same promise in the $+$ direction. We may thus proceed along the chain from the originator of the intent to the last-mile delivery proxy. Similarly, we simplify the conditional assistance notation $b(\pi_p)$ to simply $b(p)$, making the body of a promise dependency be a sufficient label for the promise itself. Then we may write the full list of promises in a compact and symmetrical form.

Figure 11.5 shows the case for three intermediate proxies, along with the construction correlations, and the promises are summarized below.

$$\text{Server} \xrightarrow{\pm S(P_1(P_2(P_3)))} \text{Client} \quad (11.31)$$
$$\text{Server} \xrightarrow{\pm S} \text{Proxy}_1 \quad (11.32)$$
$$\text{Server} \xleftarrow{\pm P_1(P_2(P_3))} \text{Proxy}_1 \quad (11.33)$$
$$\text{Proxy}_1 \xrightarrow{\pm P_1(S) \wedge (P_2(P_3))} \text{Client} \quad (11.34)$$
$$\text{Proxy}_1 \xrightarrow{\pm P_1(S)} \text{Proxy}_2 \quad (11.35)$$
$$\text{Proxy}_1 \xleftarrow{\pm P_2(P_3)} \text{Proxy}_2 \quad (11.36)$$
$$\text{Proxy}_2 \xrightarrow{\pm P_2(P_1(S)) \wedge (P_3)} \text{Client} \quad (11.37)$$
$$\text{Proxy}_2 \xrightarrow{\pm P_2(P_1(S))} \text{Proxy}_3 \quad (11.38)$$
$$\text{Proxy}_2 \xleftarrow{\pm P_3} \text{Proxy}_3 \quad (11.39)$$
$$\text{Proxy}_3 \xrightarrow{\pm P_3(P_2(P_1(S)))} \text{Client} \quad (11.40)$$

Compare (11.31) with (11.40) and observer the symmetry. Then observe how the symmetry moves along the list of intermediaries through the list from top to bottom or vice versa. Apart from promises to the client, which are always in the same direction, promises move both up and down along the chain, and each step exchanges a forward promise for a equivalent reverse promise, e.g. compare (11.32) and (11.39). The promises containing the AND (∧) symbol indicate that the promises made by the proxies to the client are conditional on *two* other promises, except for the final last-mile proxy, i.e. that they are in receipt of the good or service from upstream, and that they have obtained a promise from another proxy agent downstream to deliver in its behalf.

It is easy to see, by induction, how to generalize this case for N intermediary agents plus a client and a server. The total number of promises, i.e. edges in the graph, in this generalization would be $4N(N + 1)$ including both $\pm$ graph edges. Hence the complexity and cost of promising an assured delivery grows like N^2 in the number of intermediaries.

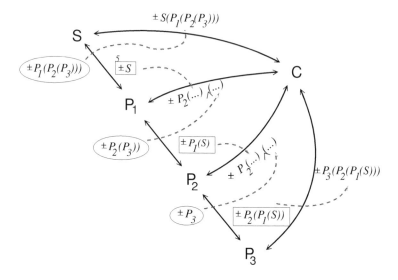

Figure 11.5: Graphical view of a three intermediary promise chain. the promises in rectangular boxes along the chain move down the chain from server S to proxy P_3 handing over the service itself. The oval promises are directed back up the chain providing delivery assurances. Finally, each agent promises independently with the end-point or client that it is an authorized agent for delivering the promise from the source (server).

It might seem surprising just how many promises are documented here. What is the difference between this and a chain of relative pushes as in section 11.4.1? In section 11.4.1, each agent merely throws a task 'over the wall' the next agent (this is sometimes called 'fire and forget'). In the approach here, each agent in the collaboration ensures continuity of its task so that errors can be repaired and the service can be maintained.

11.5. TRANSFORMATION CHAINS OR ASSEMBLY LINES

Thus this chain applies to both one-off delivery of a package and continuous delivery of a service, such as network traffic.

This example represents the extreme end of obtaining maximum certainty through signalling of intent. For continuous delivery scenarios, this represents the minimum level of assurance for a process to be fully managed.

Law 10 (End-to-end delivery assurance). *The cost of fully promised continuous end-to-end delivery grows $O(N^2)$ in the number of intermediary agents.*

The cost of an end to end assurance can be compared to the cost of $O(N)$ for the 'fire and forget' approach in section 11.4.1, with only nearest neighbour assurances.

11.4.4 END-TO-END INTEGRITY

In the worst case one could make no promises between agents and 'see what happens'. The usefulness of documenting and making these promises lies in seeing how information about agents' intentions needs to flow, and where potential points of failure might occur due to a lack of responsiveness in keeping the promises.

A chain of trust in this picture is implicit in this picture. Although the information about the promises made by $P_1, \ldots P_N$ is passed by to S through the use-promises moving backwards along the chain, S has no direct communication with them. It would not be able to detect if P_1 distorted the promise made by P_2, for example. One could add further promises from each of the proxies directly to S about how what they intend to do with the good/service they are delivering, e.g.

$$\text{Proxy}_N \xrightarrow{\pm P_N} \text{Server} \quad (11.41)$$

Then the promise to the client would be modified to

$$\text{Server} \xrightarrow{\pm S(P_1(P_2(P_3))) \wedge (P_1) \wedge (P_2) \wedge (P_3)} \text{Client} \quad (11.42)$$

In many real world situations one opts to trust because this cost of tracking and verifying becomes too much to understand and follow up.

11.5 TRANSFORMATION CHAINS OR ASSEMBLY LINES

The final generalization of the previous section is where agents do not merely pass along a service, maintaining its integrity, each agent also makes a change in the manner of an assembly line.

A typical application for this kind of model is the scenario depicted in figure 11.6, in which a product is designed and then built and distributed to customers through a delivery chain. Although we want to think in terms of this narrative, the reality of ensuring cooperation throughout is much more complicated. The model in section 11.4.3 shed some light on this case too.

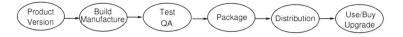

Figure 11.6: Schematic production line for goods or services.

It was assumed in foregoing sections that the proxy agents merely pass on the business of the service S unaltered, however the precise promises $\pi_{P_1} \ldots \pi_{P_N}$ have not been specified and can easily be reinterpreted without any further ado to include a transformation of what they are promised. Thus, we may use the example

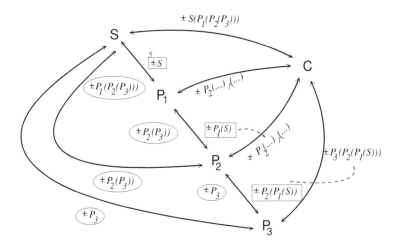

Figure 11.7: Adding direct promises to avoid deceptions and distortions along the chain.

Clearly, in the case of minimal trust, the promise graph would be a complete graph, with every agent telling every other of its intent. As intermediate agents promise radical transformations of data, trust in their promises by the point of service could become harder to swallow. Thus one can easily imagine that the most efficient delivery chains are those that make simple unambiguous promises.

CHAPTER 12

TRANSPORT NETWORK SYSTEMS

Transport networks are now familiar interconnection spaces for vehicles, electronic data, electricity, water, and several other utilities in society.

The promises made by such networks (and realized through them) lead also to networks of promises: promise graphs on top of a transport graphs, contrasting intent with information[47].

12.1 VIRTUAL CIRCUITS

The analogue of the end-to-end delivery problem for networks is the end-to-end network transport problem, which is known as a virtual circuit, as it connects a network interface at one end of a 'cloud of switches' to a network interface at the destination. See figure 12.1. The agents in a network delivery are the network interfaces themselves. These are attached to computing devices or to switches or routers. Here we shall assume that we want to make a virtual circuit or pipe from a start to a destination, via a switch that connects various other devices together. To build a promise model, we can simply apply the service via proxy in section 11.4.2, making the association:

Proxy model	Virtual circuit
Server	Start interface
Proxy	Forwarding switch
Client	End interface

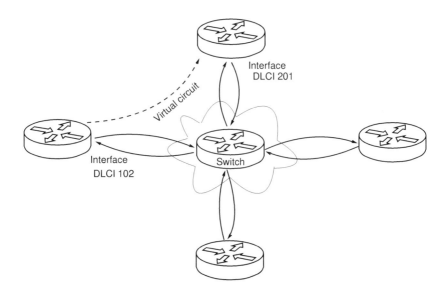

Figure 12.1: A virtual circuit, like frame relay.

Example 68. *Frame Relay is a simple example of a virtual circuit technology. End devices promise a unique name/identifier called a Data Link Connection Identifier or DLCI. Each interface is connected by a wire to a counterpart interface on a switch. The switch promises to forward traffic from one DLCI to another.*

In figure 12.1 we see two interfaces that promise DLCI identifers 102 to 201. The dotted line shows the logical pipe that represents the promise from start to end-point. The solid lines show the promises directly via the proxy or switch. There are thus three agents making the following promises:

$$\text{Interface 102} \xrightarrow{+\text{DLCI 102 if forward}} \text{Interface 201} \quad (12.1)$$
$$\text{Interface 102} \xrightarrow{\text{Use(forward 102}\rightarrow\text{201),}+\text{DLCI102}} \text{Switch} \quad (12.2)$$
$$\text{Switch} \xrightarrow{+\text{forward 102}\rightarrow\text{201}} \text{Interface 102} \quad (12.3)$$
$$\text{Switch} \xrightarrow{+\text{forward DLCI 201 if DLCI 102}} \text{Interface 201} \quad (12.4)$$
$$\text{Interface 201} \xrightarrow{\text{Use(DLCI 102 if forward)}} \text{Interface 102} \quad (12.5)$$
$$\text{Interface 201} \xrightarrow{\text{Use(forward DLCI 201 if DLCI 102)}} \text{Switch} \quad (12.6)$$

This is only half the circuit, with uni-directional forwarding from the interface that promises DLCI 102 to the interface that promises to receive as DLCI 201. We see that the Frame Relay protocol acts in a very autonomous manner. We need only basic promise types and a minimal exchange of information to pass data from end to end.

12.2 APPLICATION SERVICES

Network computer services are collaborative organizational structures based on information technology. Physical network architectures are typically built from hardware that makes the common promises described in section 4.5. They end up in many cases supporting information processes that follow the coarse architecture shown schematically in fig. 12.2.

Figure 12.2: A schematic diagram a typical multi-stage information configuration on the left, and its physical realization on the right.

We can model these common information retrieval structures as promise graphs, composed of the patterns described in section 4.5, and attempt to study their systemic behaviour. Thus we can begin to ask what it means to promise systemic behaviours like robustness, scalability, and other emergent qualities.

12.3 MULTI-COMPONENT PROMISE ARCHITECTURES

A typical information retrieval system is composed of a client talking to some kind of server array. The component agents of such a system are connected together both physically and logically, by cables and by promises. In this chapter we shall only consider the logical promise connections between these components.

Consider the three logical configurations in fig 12.3. The links or edges of these diagrams represent promises, not physical connections. The physical connections typically look more like those on the right hand side of figure 12.2. The promises represent different ways of solving the problem of connecting a client to a number servers in order to scale the service. Thus we may examine how these elementary promise configurations configurations address the systemic promise of *service capacity*, and other common promises like *response time*[48]. Let's consider the three configurations.

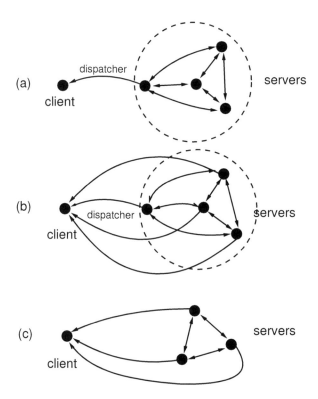

Figure 12.3: Alternative promise architectures for servicing data from a server to a client.

12.3.1 CASE (A): FRAGILE LOAD BALANCING WITH CAPACITY BOTTLENECK

The patterns used in this example are composed from the dispatcher and server-pool patterns from sections 4.5.4 and 4.5.6. A set of promises that is $\pm$ balanced unleashes a response on the arrival of an initial imposition event, or request R, from the client. This causation through the network of promises.

$$\text{Client} \xrightarrow{R} \blacksquare \quad \text{Dispatcher} \tag{12.7}$$
$$\text{Dispatcher} \xrightarrow{-R} \text{Client} \tag{12.8}$$
$$\text{Dispatcher} \xrightarrow{+R \wedge \text{select}(j)|Q_i} \text{Server}_j \in \{S_i\} \tag{12.9}$$
$$\{\text{Server}_i\} \xrightarrow{-R} \text{Dispatcher} \tag{12.10}$$
$$\{\text{Server}_i\} \xrightarrow{+Q_i} \text{Dispatcher} \tag{12.11}$$
$$\tag{12.12}$$

12.3. MULTI-COMPONENT PROMISE ARCHITECTURES 131

$$\{\text{Server}_i\} \xrightarrow{+S_i|\text{select}(i)} \text{Dispatcher} \qquad (12.13)$$

$$\text{Dispatcher} \xrightarrow{-S_i,-Q_i} \{\text{Server}_i\} \qquad (12.14)$$

$$\text{Dispatcher} \xrightarrow{+S|S_i} \text{Client} \qquad (12.15)$$

$$\{\text{Server}_i\} \xrightarrow{C(S_i)} \{\text{Server}_j\} \qquad (12.16)$$

In words, the promises are:

12.7 CLIENT IMPOSES REQUEST.
12.8 DISPATCHER PROMISES TO ACCEPT REQUESTS.
12.9 DISPATCHER PROMISES TO SELECT A SERVER FROM THE POOL.
IF IT KNOWS THE SERVER QUEUE LENGTHS Q_i.
12.10 SERVERS PROMISE TO ACCEPT REQUESTS FROM THE DISPATCHER.
12.11 SERVERS PROMISE TO REPORT ON THEIR LOAD OR QUEUE LENGTHS Q_i.
12.13 SERVERS PROMISE TO REPLY TO REQUESTS WITH A SERVICE RESPONSE
TO THE DISPATCHER.
12.14 DISPATCHER PROMISES TO ACCEPT SERVER RESPONSES AND QUEUE INFO.
12.15 DISPATCHER PROMISES TO FORWARD REPLIES BACK TO THE CLIENT.
12.16 SERVERS PROMISE TO DELIVERY THE SAME SERVICE CONTENT.

What systemic promises can be inferred from these underlying cooperative promises? The presence of multiple agents in the role of server implies that the capacity for work is greater than for a single server. Internally, the work must be proportional to the number of servers, but outside the service capacity is limited by the capacity of the smallest bottleneck [49], which might be the selected server or the dispatcher.

- The collaborative superagent formed by the cooperation between servers and dispatcher can thus make promises about service capacity. The expected capacity will be

$$\{\text{Server}_i, \text{Dispatcher}\} \xrightarrow{\min(\sum_i \text{capac}(S_i)),D} \text{Client} \qquad (12.17)$$

- Similarly, the latency or delay in response is limited by the server and two passes through the dispatcher agent:

$$\{\text{Server}_i, \text{Dispatcher}\} \xrightarrow{\text{delay}(S_i+2D)} \text{Client} \qquad (12.18)$$

These are the macro properties of the system of collective agent promises.

If the promises are not made explicitly by some representative agent, the ability of external agents to reason the existence of implicit promises, based on the internal ones, depends on the scope of the internal promises, i.e. whether they are visible to outside agents or not.

Super agents like this pay the role analogous to subroutines in computer programming. They often expose certain promises (called an Application Programmer Interface or API) and the internal details are out of scope to outsiders. This is essentially the role of the dispatcher here.

12.3.2 Case (B): Fragile load balancing with queue bottleneck

In the second architecture, the dispatcher does not act as a go-between for replying to clients. Instead it merely delegates to one of the servers which promises to reply directly to the client[50].

$$\text{Client} \xrightarrow{R} \text{Dispatcher} \quad (12.19)$$
$$\text{Dispatcher} \xrightarrow{-R} \text{Client} \quad (12.20)$$
$$\text{Dispatcher} \xrightarrow{+R \wedge \text{select}(j)|Q_i} \text{Server}_j \in \{S_i\} \quad (12.21)$$
$$\{\text{Server}_i\} \xrightarrow{-R} \text{Dispatcher} \quad (12.22)$$
$$\{\text{Server}_i\} \xrightarrow{+Q_i} \text{Dispatcher} \quad (12.23)$$
$$\text{Dispatcher} \xrightarrow{-Q_i} \{\text{Server}_i\} \quad (12.24)$$
$$\{\text{Server}_i\} \xrightarrow{+S_i|\text{select}(i)} \text{Client} \quad (12.25)$$
$$\{\text{Server}_i\} \xrightarrow{C(S_i)} \{\text{Server}_j\} \quad (12.26)$$

The promises are now:

12.19 CLIENT IMPOSES REQUEST.
12.20 DISPATCHER PROMISES TO ACCEPT REQUESTS.
12.21 DISPATCHER PROMISES TO SELECT A SERVER FROM THE POOL. IF IT KNOWS THE SERVER QUEUE LENGTHS Q_i.
12.22 SERVERS PROMISE TO ACCEPT REQUESTS FROM THE DISPATCHER.
12.23 SERVERS PROMISE TO REPORT ON THEIR LOAD OR QUEUE LENGTHS Q_i
12.24 DISPATCHER PROMISES TO ACCEPT QUEUE LENGTH INFORMATION FROM THE SERVERS.
12.25 SERVERS PROMISE TO REPLY TO REQUESTS WITH A SERVICE RESPONSE DIRECTLY TO THE CLIENTS.
12.26 SERVERS PROMISE TO DELIVERY THE SAME SERVICE CONTENT.

12.3. MULTI-COMPONENT PROMISE ARCHITECTURES

This leads to a modification of the expected capacity, since the capacity is no longer throttled by the need for the dispatcher to keep a promise.

$$\{\text{Server}_i, \text{Dispatcher}\} \xrightarrow{\sum_i \text{capac}(S_i)} \text{Client} \quad (12.27)$$

$$\{\text{Server}_i, \text{Dispatcher}\} \xrightarrow{\text{delay}(S_i+D)} \text{Client} \quad (12.28)$$

12.3.3 CASE (C): ROBUST LOAD BALANCING WITH NO BOTTLENECK

In the final diagram from figure 12.3, we remove the dispatcher altogether. By making the client itself the dispatcher, we cut out the middle man[51] and the client merely chooses a server itself. This is analogous to shopping around for your own deal rather than going to a broker. The promises are now:

$$\text{Client} \xrightarrow{R} \text{Server}_i \quad (12.29)$$

$$\text{Client} \xrightarrow{-Q_i} \{\text{Server}_i\} \quad (12.30)$$

$$\{\text{Server}_i\} \xrightarrow{-R} \text{Client} \quad (12.31)$$

$$\{\text{Server}_i\} \xrightarrow{+Q_i} \text{Client} \quad (12.32)$$

$$\{\text{Server}_i\} \xrightarrow{+S_i|R} \text{Client} \quad (12.33)$$

$$\{\text{Server}_i\} \xrightarrow{C(S_i)} \{\text{Server}_j\} \quad (12.34)$$

In words:

12.29 CLIENT IMPOSES REQUEST.
12.30 CLIENT PROMISES TO ACCEPT QUEUEING INFORMATION FROM THE SERVER.
12.31 SERVERS PROMISE TO ACCEPT REQUESTS FROM THE CLIENTS.
12.32 SERVERS PROMISE TO REPORT ON THEIR LOAD OR QUEUE LENGTHS Q_i
12.33 SERVERS PROMISE TO REPLY TO REQUESTS WITH A SERVICE RESPONSE DIRECTLY TO THE CLIENTS.
12.34 SERVERS PROMISE TO DELIVERY THE SAME SERVICE CONTENT.

The capacity and delays one may promise now become very clear:

$$\{\text{Server}_i\} \xrightarrow{\sum_i \text{capac}(S_i)} \text{Client} \quad (12.35)$$

$$\{\text{Server}_i\} \xrightarrow{\text{delay}(S_i)} \text{Client} \quad (12.36)$$

12.3.4 REMARKS ABOUT SYSTEMIC PROMISES

One observes that the introduction of an intermediate agent in the network leads to an increase in the complexity of assuring behaviour, and potentially limits the capabilities that can be promised. The absence of a dispatcher role, in these examples, simplifies the system greatly, and makes it possible for the client to exploit all of the servers' promises in parallel, not merely one at a time. With this architecture, there is no single point of contact, so extra information is required by the clients to locate the services.

The number of promises used to seek assurance in collaborative systems can be used as a rough measure of system complexity. Each promise documents a possible failure mode, and hence one can also test the fragility of the system.

Finally, we note that a so-called single point of failure occurs when the failure of a single promise by a single agent can cause the system in total to fail. One observes from figure 12.3 that putting a dispatcher in front of a redundant ensemble of service agents merely pushes the fragility of a single server back to a single dispatcher promise, which might increase availability but not robustness. The third solution pushes the problem all the way back to the client, where it becomes unimportant: if the client fails, it cannot make a service request.

An alternative solution might be to introduce redundancy into the dispatcher agent's role, and duplicating the promises to provide dispatching to all server agents (see figure 12.4). This additional robustness comes at the cost of $O(N^2)$ complexity to the system.

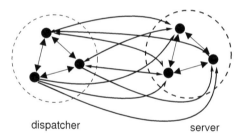

Figure 12.4: A robustness leads to strong coupling, which merely shifts fragility to a larger scale.

The lesson here is clearly that intermediate agents are expensive to introduce and lead to complexity.

CHAPTER 13

SYSTEMIC PROMISES

Promises come from agents that are the indivisible building blocks of behaviour: the lowest level entities in a system, like atoms in a chemistry of intent. These agencies of behaviour may be chosen with great freedom when modelling systems. Sometimes we draw a boundary around a country, sometimes just around a person, depending on what scales we wish to highlight.

A *system* is a collection of parts that work together in order to carry out a collective function. Systemic promises describe qualities we expect from the collective behaviour. One associates the concept of *organization* with systems in two senses: organization as an entity (also called institution) and as a state of being (an orderly condition).

In this chapter we consider how we can make high level promises about the *systemic* characteristics of agents, and how these may be related to the low level promises of the elementary parts through cooperative behaviour.

13.1 AGENT TYPE DIFFERENTIATION

Let us now examine how modular *discrimination* enables *order*. Is a tree structure considered organized, or merely ordered? The (by now) well established term self-organization forces us to define the meaning of organization clearly, since it implies that organization may be something that is both identified *a priori* by design, or *a postiori* as a system property.

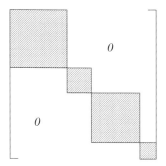

Figure 13.1: Patterns in agent space arise from irreducibility of the graph.

13.1.1 ORGANIZATION OF AGENTS INTO ORDERED STATES

Intuitively we think of organization to mean the tidy deployment of resources into a structural pattern. "Organization" (from the Greek word for tool or instrument) implies to us a compartmentalization of roles, i.e. a *modularity*[52]. Organization requires distinguishability of agents.

> **Definition 46** (Organization). *A phenomenon in which a pattern forms in the behaviour of an ensemble of differentiated agents.*

The promise graph is a spatio-temporal structure: some promises in the graph become defined at different times, connecting agents temporarily. Patterns may be formed over promises in both space and time.

- Spatial or role-based partitioning of operations between parallel agents.

- Temporal (schedule), i.e. serial ordering of operations at an agent.

More generally, orthogonal types of organization are represented by longitudinal and transverse patterns:

> **Definition 47** (Serial (longitudinal) organization). *A sequence of promises that follow one another, represented as a path through the promise graph.*

> **Definition 48** (Parallel (transverse) organization). *A partitioning of pathways through a promise graph, induced as irreducible modules of the promise graph each serial step*[53].

13.1. AGENT TYPE DIFFERENTIATION

13.1.2 SPECIALIZATION AND FUNCTIONAL REUSABILITY

Specialization into patterns representing roles can also lead to the systemic property of reusability (see section 14.5). Like promises, stable roles form for semantic reasons, and persist for economic or dynamical reasons.

Definition 49 (Specialization and separation of concerns). *A 'division of labour' in the structure of a promise pattern, whereby agents specialize around keeping certain promises and collaborate with other agents that have different roles.*

Example 69. *Two agents trained to fight a fire could both independently promise to grab the fire extinguisher or dial 911, but if they promise to divide the tasks then both tasks will be started sooner and finished earlier costing less totally and improving efficiency by parallelism.*

The economic efficiency gain is the seed for *differentiation*; its survival is a matter of sustained advantage, which requires sustained environmental conditions.

13.1.3 ORGANIZATIONS OR INSTITUTIONS

An organization is an ensemble with collective behaviour. It forms a pattern, even if only a trivial one; however, organizations or institutions as we understand them always have boundaries. One may or may not consider these artificial, but they exist. Organizations interact with other agents outside their boundary, so the web of promises extends beyond the boundary, making it difficult to define an organization by its functional promises alone[54].

Any ad hoc definition of the edge of the organization as the edge of the pattern would be arbitrary, and a matter for individual assessment. No agent would be able to know whether it were inside or outside the organization itself in an agreed sense. However, here lies the clue: how would an agent know if it belonged? The boundary is a promise of membership. The common meaning of an organization is as follows:

Definition 50 ('An' organization). *A number of agents that each promises to be identified as members of an organization.*

Example 70. *Uniforms and membership cards are promises given to potential members of an organization. Accepting these symbols of membership constitutes a use-promise that leads to institutional binding.*

Organizations are thus more than a pattern that identifies 'collective agents' making promises that a single agent would not be able to make alone. They are also mutually-appointed *roles*.

13.2 SUPERAGENTS: AGENTS MADE OF OTHER AGENTS

How we choose to model a system in terms of agents depends on what level of detail we wish to expose about the promises of a system.

Example 71. *If we are interested in traffic, we could consider an entire car to be an indivisible agent in a promise model. However, if we later need to compare cars, finer granularity is needed. Then the agents could be wheels, doors, chassis, seats, etc, each of which make specific promises. Each of these agents (which can exist independently in the factory) must then promise to play roles in belonging to the same car in order to be considered an organized entity.*

Using the concept of an organization, we may thus define the notion of an agent that is composed of other agents, working together under a common label. This is simply an ensemble of agents that fulfill a particular role.

When the ensemble of agents makes a promise, it must transfer to subsets of agents from within the ensemble, as a role has not physical existence except in terms of the agents that comprise it.

- A single point of entry, or role representative, to the ensemble (see section 4.5.6) that agree to work together (see section 8.4.4).

- A number of agents that make collaborative promises to collectively represent the outward promise (see section 6.3).

See the example in section 12.

To describe the composition of agents, we need to describe how agency scales collectively, through aggregation and reduction of elementary agents[Bur15]. The concepts of scaling are familiar to physicists for dynamics, but here we also want to extend them to incorporate semantics. This might be motivated by questions of the following kind:

- How does a team promise something as a unit?

- How does an organization appear as a coherent entity?

- How does a collection of components promise to be a car?

13.2. SUPERAGENTS: AGENTS MADE OF OTHER AGENTS

Since we can aggregate several promises into a single promise, and aggregate agents into a single agent, the ability to detect or resolve parts within a whole depends on the observer's capabilities. Similarly, the ability for an agent to perceive a collection of individual agents with a collective identity (i.e. a superagent) depends on the capabilities of the observer agent. Elementarity and composability of agency thus go together with a hierarchy of observable agency, which needs to be elucidated.

13.2.1 COMPOSITION OF AGENTS (SUB AND SUPERAGENCY)

The treatment of a collection of agents as a single entity is a choice made by any observer. It can be made with or without promises from the composite agents themselves (see figure 13.2). Agency can be defined recursively to build up hierarchies of component parts. In [Bur14], it was shown how spatial boundaries can be defined by membership to a group or role. We still have to explicate the relationship between the internal members and the structure of the whole, as perceived by an observer.

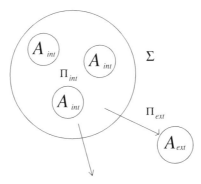

Figure 13.2: Agent structure consists of an element that makes a number of exterior promises, some of which are scalar, some vector, etc. Interior promises are invisible from the outside.

We define a collective superagent as a spacetime structure that has collective agency, i.e. its intended semantics relate to a collection of agents surrounded by a logical boundary, with collective semantics (see figure 13.2).

Definition 51 (Bare superagent). *A superagent of size S is any bounded agency composed of individually separable agencies, partially or completely linked by internal vector promises. The bare superagent is defined by the closed graph, without any external adjacencies. It is a doublet:*

$$A_{\text{super}} = \langle \{A_i\}, \Pi_{ij}^{\text{int}} \rangle, \quad i = 1, 2 \ldots S. \tag{13.1}$$

where A_i is an internal agent of A_{super}, and Π_{ij} is the promise adjacency matrix between the S internal constituents.

Definition 52 (Dressed superagent). *A dressed superagent is the bare superagent together with its set of exterior promises. It is a triplet:*

$$A_{\text{super}} = \langle \{A_i\}, \Pi_{ij}^{\text{int}}, \Pi_{s \epsilon 1}^{\text{ext}} \rangle, \quad i = 1, 2 \ldots S. \tag{13.2}$$

Superagency allows promises to exist within and without a superagent boundary. We call these interior and exterior promises, respectively, and define them as follows:

Definition 53 (Exterior promises). *Exterior promises are made by agencies within the superagent boundary, to agents outside. They represent inputs and outputs of the superagent, i.e. how it interacts with an external space.*

Definition 54 (Interior promises). *Internal promises are made by agents inside the superagent boundary to other agents inside the boundary. They represent the bindings that make the superagent behave as a single cohesive entity.*

In principle an observer could draw a line around any collection of agents and call it a cell or composite superagent. This is an assessment any agent can make, as part of its definition of agent scale. However, it might still be of interest to distinguish special criteria by which such an arbitration might occur. A privileged arbiter is one possibility. In component design, for instance, boundaries may be chosen on the basis of the specific interface an agent exposes or interacts with.

Superagents are formed by interior connectivity of the member agents. Alternative structures fall into three basic categories (see figure 13.3):

(a) A membership in a group or associative role, where the central membership authority may be either inside or outside the boundary. In this case, we are identifying a group of symmetrical agents.

$$A_{\text{host}} \xrightarrow{+\text{membership}} \{A_{\text{tenant}}\} \tag{13.3}$$

$$\{A_{\text{tenant}}\} \xrightarrow{-\text{membership}} A_{\text{host}} \tag{13.4}$$

13.2. SUPERAGENTS: AGENTS MADE OF OTHER AGENTS

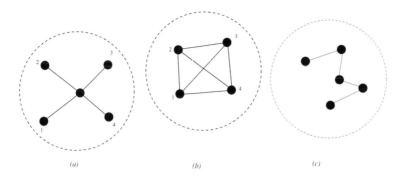

Figure 13.3: Three ways of binding collective agency. Another way is to simply make an arbitrary collection.

(b) A total graph or collaborative role. In this case, we are identifying agents with coordinated behaviours.

$$A_i \xrightarrow{\pm \text{membership}} A_j \quad \forall A_i, A_j \in A_{\text{super}}. \qquad (13.5)$$

(c) A dependency graph, path or story. In this case we are identifying dependency bindings.

Let us now define the converse properties:

Definition 55 (Subagent). *A subagent is an agent assessed to be a resident of the internal structure forming a composite (super) agent.*

Definition 56 (Residency). *A subagent A is resident at a location L iff it is defined to be within the boundary of the agent:*

$$A \cap L \neq \emptyset. \qquad (13.6)$$

Since an observer can form their own judgement about superagent boundaries, we cannot say that residence is the same as a promise of adjacency.

There are two types of adjacency, somewhat spacelike, which may or may not be interchangeable (see figure 13.4). Normal 'physical' adjacency promises, and resident adjacency, which might link agents virtually even though they are not physically adjacent. I'll return to this topic when discussing tenancy below.

Lesser agents can become satellites of other agents. This leads to a hierarchy or 'planetary' structure, by accretion into superagents.

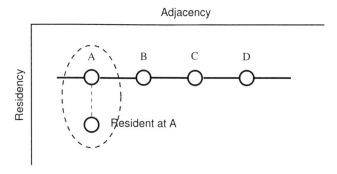

Figure 13.4: A resident adjacency forms a superagent by accreting to a seed agency that represents the location anchor.

13.2.2 SUPERAGENT SURFACE BOUNDARY

superagent boundaries may be formed with different structural biases.

- Simple aggregations of agents, related through membership to a single leader (leader may be inside or outside the superagent), and the connections are made through the leader as a proxy-hub.

- A cluster of agents linked by cooperative vector promises.

- Strongly cooperative agents which are inseparable without breaking an external promise. E.g. an organism made of components that are all different and non-redundant to the functioning of the whole.

Interior promises are those entirely within the surface boundary of a superagent. We may define interior and exterior promise matrices for any agency, using a matrix analogous to the adjacency matrix:

$$\Pi^{\text{int}}_{ij} = \left.\begin{array}{l} 1 \quad \text{iff } A_i \xrightarrow{*} A_j \\ 0 \end{array}\right\} \quad A_i, A_j \in A_{\text{super}} \quad (13.7)$$

$$\Pi^{\text{ext}}_{i\epsilon} = \left.\begin{array}{l} 1 \quad \text{iff } A_i \xrightarrow{*} A_\epsilon \\ 0 \end{array}\right\} \quad A_i \in A_{\text{super}}, A_\epsilon \notin A_{\text{super}} \quad (13.8)$$

Definition 57 (Surface of a superagent). *The exposed surface Σ of the agent is the subset of interior/internal agents that have adjacencies to agencies outside the superagent.*

$$\Sigma \equiv \{A_i\} \subset A_{\text{super}} \;\Big|\; \Pi^{\text{ext}}_{i\epsilon} \neq 0 \quad (13.9)$$

13.2. SUPERAGENTS: AGENTS MADE OF OTHER AGENTS

A superagent surface may also make new explicit promises that are not identifiable with a single component agency (see section 13.2.3).

Example 72. *In the molecular example above, the superagent makes interior promises*[55]

$$A_1 \xrightarrow{+H} * \quad (13.10)$$

$$A_2 \xrightarrow{+H} * \quad (13.11)$$

$$A_3 \xrightarrow{+O} * \quad (13.12)$$

from internal agents, and collectively $M = \{A_1, A_2, A_3\}$

$$M \xrightarrow{+H_2O} * \quad (13.13)$$

This promise implies some interior structure, and it does not emanate from any smaller agent. Thus if an external observer is able to resolve the component agents within M, the promise of H_2O is not longer a promise.

An observing agent may or may not be able to discern an internal elementary agent within a superagent, i.e. whether the agent has internal structure, or whether it is atomic. This depends on whether the agent promises transparency across its surface.

Definition 58 (superagent transparency). *All promises whose scope extends beyond the boundary pass transparently through the surface of the superagent. Scope includes the list of promisees.*

13.2.3 IRREDUCIBLE PROMISES AT SCALE M, AND COLLECTIVE BEHAVIOUR

In addition to the exterior promises, which emanate from composite superagencies, as the remnants of microscopic promises belonging to component subagencies, we may also observe that completely new promises are possible at each scale, which do not belong to any specific agent inside the surface.

Definition 59 (Irreducible superagent promises). *Let M be an agency scale, and A_s be a superagent formed by an aggregation of agents. A promise with body b_s made by A_s*

$$\pi_M : A_s \xrightarrow{b_s} A_? \quad (13.14)$$

$$A_s = \{A_i, \ldots\} \quad (13.15)$$

may be called irreducible iff there is no set of subagents $A_i \in A_s$, for which

$$b_s \subset \bigcup_a b_a, \quad a = 1, \ldots \text{ all promises to } A_? \quad (13.16)$$

where b_a are existing bodies promised to $A_?$ by A_i, and $A_i \neq A_s$.

In other words, if there exists a combination of promises made by one or more subagents (and we assume that the subagent is not the same as the superagent), then that is semantically equivalent to the full promise made by the composite agent, when one could say that the superagent promise could be reduced to the promise of one of its components. As long as no single agent, working alone, can make such a promise, it makes sense to talk about the collective superagency making a new promise that is not explicit in the capabilities of its subagencies. Thus, irreducible promises at scale M take into account emergent effects, and collective effects of agent interactions.

Example 73. *A radio is a composite agent that makes the exterior promise of playing radio signals on a loudspeaker. Interior promises are made by many component agents like resistors, capacitors, transistors, etc. The radio has a semantic surface which interacts with people as promisees[56]. Does this mean that the components inside the radio have to interact with the components inside a person (cells, organs etc)? Not really. Implicitly, this might be partially true by the distributive rule, but clearly that is not a requirement. Any agency can interact, semantically, with any other agency at any agency scale, provided there is a physical channel for the communication to work.*

Irreducibility is thus a result of collective phenomena in the underlying semantics and dynamics. The latter answers classic objections against the naive reductionism of systems. Yes, we can decompose systems into a sum of parts, but one must not throw away the promises when doing do, else it will not be possible to reconstruct both semantics and dynamics. This is a consequence of boundary information or systemic topology.

The form or collective identity of a superagent can be enough to signal a promise, by association. This is an emergent promise i.e. one inferred by a potential promisee rather than given explicitly by an agency. Nevertheless, some scaled objects do make promises: tables, chairs

Law 11 (Reduction law). *When reconstructing a system from components from finer-grained scale M' to a scale M, all superagencies at scales below M, and their component promises must be retained in order to reconstruct the system as the sum of its parts. This is achieved if each superagent promises its coarse-graining directory, providing dynamic transparency.*

Thus increase of detail downwards may not be at the expense of loss of upward irreducible information. If we try to view a compound agent as a collection of parts, with component promises (on the interior of the superagent), the component promises revealed do not replace the high level exterior promise: they are simply prerequisites for it. The only reason to disregard irreducible promises belonging to a coarse grain is because one is

13.2. SUPERAGENTS: AGENTS MADE OF OTHER AGENTS

focusing on the internals of an agent in isolation (what one calls a closed system in physics).

If a superagent has no agency of its own, how can it make a promise? Clearly, a collection of agents has agency through its members, and these may or may not have the ability to keep promises related to scale M. So where do promises come from in a superagent?

- $+$ promises: the collective appearance of a superagent, at a certain scale, must provide the information to signal a promise. The declaration of the promise may or may not come from a single agent. The keeping of an irreducible promise does not come from a single agent, by definition. New irreducible promises are dependent on the individual agents, only through their cooperative behaviours.

    Example 74. *A troop unit promises to surround a house (+ promise). The troop leader can make the promise on behalf of the group, but no single agent can keep this promise, but collectively they can. In this case, the promise would often be given by a team commander, with a centralized source of intent, and subordinate agents. However, a team can also arrive at this promise by cooperative consensus.*

- $-$ promises: a promise to accept another promise, made by an external agent, might only apply to certain subagents with the appropriate capabilities. It must be provided as an exterior promise based on the by cooperative agreement amongst the agents.

    Example 75. *The Very Large Array of radio telescopes in Mexico has 27 receivers. Each receiver (- promise) is coordinated with the others so that they act as a single superagent. No single agent can see what the full array can see working together, up to diffraction. Hence the combined array can make promises that individual agents can't.*

A superagent's *agency* is thus conditional on the existence of its subagencies, even though its promises are not locatable in any single agent. In both cases above, the promise made by a superagent could not be made by a single agency (this is what we mean by a 'host'), or by distributed consensus. Irreducible promises are thus conditional, not only on the uniform cooperation of the subagents about a single promise, but on their making all the promises that indirectly lead to the irreducible property.

Lemma 7 (Irreducible promises are conditional promises, for all scales greater than M_0). *Irreducibility is not expressible directly as a sum of component promises of the same type as the irreducible promise, but it is second-order expressible in terms of subagent promises, by building on the existence of these contributing promises.*

With no promises to build on, a collective agent cannot make any kind of promise, since nothing can be communicated, and it has no independent agency. Crudely, a superagent is indeed the sum of its parts, as far as agency and promise-keeping are concerned; but, it is not merely a direct sum as new promises are possible through cooperation.

If the scope of exterior promises extends to the interior of a superagent boundary, that scope becomes ambiguous under coarse-graining. External observers can no longer see which agents are make or receive the interior promises, nor is there any way to refer to them independently after coarse-graining. Observers can only assume that the scope of a promise includes all subagents, but this might not be the case, and might result in erroneous expectations. Indeed, there is no reason why the subagents would even all use the same body language.

13.3 Centralization, decentralization, and hierarchy

The role played by centralization versus decentralization in systems is a complex topic that has great emotional resonance with readers. Most people's ideas about this topic are forged on the anvil of organizational structures in life, e.g. at work. For example, centralized government versus decentralized power. Government institutions versus market forces. The language used to describe these topics is often loaded with connotations about which system we are supposed to prefer, if we are good humans. For example, conservatives and libertarians tend to reject central government on ideological grounds, preferring market forces as a mechanism for resolution of challenges. Authoritarians may advocate for centralized control from the top down. Leftists may advocate for shared ownership or a free commons. These caricatures do not do justice to the topics by any means. There are complex scaling issues involved in finding optimal cooperative models, and usually a partially (de)centralized compromise solution is the only way to avoid certain problems. The problem with extreme decentralization is that it is free to self-organize and fall into traps called 'local minima' that are sub-optimal, and which may fragment purpose into cellular regions, counter to intent; however, its effort scales unconstrained without limit as long as no agent needs another. The problem with centralization is that it is fragile as a bottleneck and may scale poorly; however, it is the most effective way to maintain a singular coherent purpose. We refer readers to [Bur19a] for more on these topics.

A structural compromise, which interpolates between centralization and decentralization, is the hierarchical model. A hierarchy needn't be precisely a tree, but its communication and promise structure will often be somewhat treelike, with a root at the the 'top' and leaves at the bottom (something like a house of cards, and with attendant

13.3. CENTRALIZATION, DECENTRALIZATION, AND HIERARCHY

fragilities).

In human systems, the concept of organization also imbues a conscious decision amongst a number of agents to work together, with possibly a hierarchical structure, and a leader. Even non-human systems may benefit from this architecture, however. In organizational literature, hierarchy lies behind practically all visions of what constitutes systematic and organized behaviour. It's hard to escape the assumption that hierarchy is both necessary and sufficient. This is a gross simplification, but with a grain of truth to it.

In line with many previous authors' thinking, we shall assume that hierarchies emerge for microeconomic reasons. We might diverge from other authors in the suggestion that this is inevitable. Each agent, after all, pursues its own interests autonomously and the resulting collective behaviour reflects an evolutionary process[Axe97, Axe84]. Despite its widespread dominance of hierarchies empirically, we resist the *assumption* of hierarchy as a pattern of organization.

A hierarchy stems from the specialized service pattern, in which a single agent takes on a specialization so that not every agent has to have that interior capability. If all N agents in a cluster had to promise the same promises, it would lead to promises between every pair of agents $O(N^2)$, rather than just with a single agent $O(N)$ with multivalent promise. So, there is a potential cost-saving advantage, and a potential risk of a resource bottleneck by creating specialized roles. Nonetheless, the pattern begins cheaply and rationally before it possibly exceeds its maximum capacity. If replicated in a chain of service dependencies, it leads to a hierarchy.

The characteristic of hierarchy is the existence of a 'top' or root node, i.e. a characteristic agent around which a cluster of other agents (and sub-clusters) forms. The question is how this node gets selected from a group of agents[57]. This requires a dynamical, i.e. economic explanation.

There may be an economic advantage to electing such a single agent to coordinate and calibrate the behaviour of a cluster. For even a handful or agents this is cheaper ($O(N)$) than having each individual agent establish peer-to-peer communication with every other agent in the ensemble ($O(N^2)$). The cost-benefit of such centralization depends on how many promises need to be set up and maintained. Two separate economic issues govern ensembles:

- The cost of *calibration* or standardization, requiring global consistency.

- The cost of *coordination* or differentiation and delegation which requires only local consistency.

Calibration requires complete bi-directional communication between all agents[58]. Coordination requires only that we can pass a message to every agent on a need to know basis, without the necessity for reply. Without calibration, agents have only local concerns and global ones are considered to 'emerge' i.e. they are un-calibrated (we return to emergent behaviour below).

Hierarchies thus emerge, in a promise view, from the structural accretion of agents under economic constraints that promise to work cooperatively.

13.4 Occupancy and tenancy of space

Let's now turn to a different topic: how to fill the space we've built up. A semantic space is richer in structure than its underlying connective graph so it contains information that goes beyond pure adjacency. In particular, as we add autonomous observers with their own agency, we quickly arrive at the need for agents to extend their realm of autonomous control through *occupancy* and *ownership* of resources. The question of occupancy and tenancy are thus about how we draw the boundaries of agency on a background of spatial adjacency So far we've focused on symmetry and scale in discussing agency, however strong functional semantics are a result of asymmetry, hence we must now pursue the effects of broken symmetry.

13.4.1 Definitions of occupancy and tenancy

Tenancy goes beyond simple aggregate membership in a cluster. A tenant is understood to be an agent that 'occupies' or utilizes a resource or service, provided by a host, often in a temporary manner, and for mutual benefit (symbiosis). Tenants have separate identities. When we think of tenancy in every day affairs, we do not usually imagine a tenant as merging with its host, and becoming a part of it (though merger and acquisition is certainly a process one can discuss, as absorption). Tenancy is rather an association between two separate agency roles (host and tenant), each of which retains its autonomy.

To relate this to our spacetime discussion, consider the following question (which, at first glance, might seem purely facetious): *does a suit occupy space when no one is wearing it, or does space occupy the suit?* The space inside a suit is simply empty before someone climbs into it. Try replacing 'suit' with 'car' and 'wear' with 'sit inside'.

This peculiar question is closely related to the considerations surrounding the kinds of motion described in paper I, section 5.12. Since we are modelling space as a resource, this is not only a meaningful question, it is essential to understand what kind of volume a suit occupies. Does the presence of suit matter replace space, occupy it, attach to it, or overlap with it? These have different semantics.

Recall, from paper I, that motion of the second and third kinds distinguish between the idea that space and its occupants are either: ii) a visitation by a separate entity, or iii) a change in the state of the same entity. In other words, does space get filled by matter or

13.4. OCCUPANCY AND TENANCY OF SPACE

does matter transform the nature of space?

To address some of these issues, we need to formulate definitions using promises, building up the distinctions in a rational way. Let's begin with occupancy. Its semantics are difference from mere presence, as there is an assumption of valency. Within the scope of Promise Theory, we can define the following:

Definition 60 (Occupancy). *An asymmetric association of one agent (the occupier) with another representing a host (the location) at agency scale M, in which the valence of a promise made by the host is reduced by its binding to the occupier. The resource R may be any scalar, vector or tensor type:*

$$\text{Host}_M \xrightarrow{+R\#n} A_? \quad (13.17)$$
$$\text{Occupier}_M \xrightarrow{-R\#1} \text{Host}_M \quad (13.18)$$

In other words, a host makes a finite promise $+R$ to a number of agents in scope, and each occupier reduces the valency by making a use-promise $-R$

$$\textbf{Valence}(R, \text{Host}) = \textbf{Valence}(R, \text{Host}, \text{Occupier}) + 1 \quad (13.19)$$

Example 76. *Our understanding of the semantics of occupancy has many possible interpretations. Here are some examples:*

- *Occupation of a place or territory without necessarily being there. e.g. a table reservation, or planting a flag on the Moon.*

- *Occupation of space.*

- *Occupy a car, a suit, a dress.*

- *Occupy a time slot in a calendar.*

- *Filling a space with something.*

- *In physics, bosons can occupy the same space, like voices in a song, but Fermions have exclusion, like the bodies in the choir themselves they occupy space.*

From here, we may state a basic template for tenancy for application to a variety of special cases:

> **Definition 61** (Tenancy). *Tenancy refers to the conditional occupancy of a location, by an agent, together with the provision of one or more services by the host, which may be considered a function $f(R)$ of resource R. These services are provided conditionally on a promise of C from the tenant:*
>
> $$\begin{aligned} \text{Host}_M &\xrightarrow{+R\#n|C} A_? \\ \text{Host}_M &\xrightarrow{-C} \text{Tenant}_M \\ \text{Tenant}_M &\xrightarrow{+C} \text{Host}_M \\ \text{Tenant}_M &\xrightarrow{-R\#1} \text{Host}_M \\ \text{Host}_M &\xrightarrow{+f(C,R)|-R} \text{Tenant} \end{aligned} \quad (13.20)$$
>
> *This is the basic template for tenancy, which may be extended by additional promises.*

As defined, every tenant is an occupant but not every occupant is a tenant (e.g. squatter).

Example 77. *A landlord promises a rentable space for a single occupant $+R\#1$, conditionally on the signing a contract of terms (i.e. the promise to abide by terms and conditions) $+C$.*

$$L \xrightarrow{+R\#1, f(C,R)\,|\,C} A_? \quad (13.21)$$

A tenant quenches this exclusive resource, by signing up and promising the terms:

$$T \xrightarrow{+C,-R} L. \quad (13.22)$$

The terms and conditions contain a composite promise body, detailing the services $f(C, R)$ offered as part of the promise:

$$C = \{+\text{payment}, \text{termination date}, \ldots\} \quad (13.23)$$

and

$$f(C, R) = \{+\text{power}, +\text{heating}, \ldots\} \quad (13.24)$$

Tenancy is a service-like relationship between a host and a tenant. This may be contrasted with the notion of residency at a location, which is related to definition of boundaries within an observer's realm. Tenancy is a also relative concept (relative to promise semantics).

13.4.2 Laws of Tenancy Semantics

It is basic to Promise Theory that we distinguish between a promise made by an agent, and the agency itself. Hence, we begin by noting that:

13.4. OCCUPANCY AND TENANCY OF SPACE

Assumption 1. *Promises are neither occupants nor tenants of the promisers or promisees, since they have no independent agency.*

- *Tenancy and occupancy requires two agencies to become associated.*

- *Agents can be promised, but promises are not agents (they do not possess independent agency).*

In a sense, a promise emanating from an agent seems to be attached at the location represented by the agent. However, we do not call this tenancy. A promise is a property of an agent, but it has no independent agency, thus it cannot be a tenant.

Example 78. *An agent A can be the subject of a promise, e.g.*

$$A_1 \xrightarrow{+A} A_2, \tag{13.25}$$

but it is not the promise itself, which belongs to A_1.

The semantics of the promises in (13.20) select an inherent directionality for the provision and use of a resource.

Lemma 8 (Tenancy flows in the direction of the resource being used). *Tenancy flows towards the host, i.e. towards to source of the hosted resource.*

It is important to bear in mind the semantics when looking at host and tenant. Consider the following case in figure 13.5. From the perspective of a renter going directly to a hosting apartment block, the tenant

Definition 62 (The host:tenant binding is 1:N). *A host can have any number tenants, at any one time, keeping full promises, up to and including the valency of the host resource promise.*

There is an exclusivity between a tenant and a resource, which is a question of definition. Tenancy with a superagent scales like any other promise (see section 13.4.8). When we speak of a tenancy, it refers to a single relationship, even though an agent might be engaged in multiple similar tenancies.

Example 79. *A rider on a horse is a tenant of the horse. A rider cannot ride a herd of horses, at the same time. Moreover, the rider and horse are not joined by physical encapsulation, forming an exterior agent, one inside the other. A driver in a car however, is a tenant of the car, and is encapsulated by it. The car is a tenant of the driver's direction. Hence, while the two have independent agency, they seem to form an encapsulated superagent.*

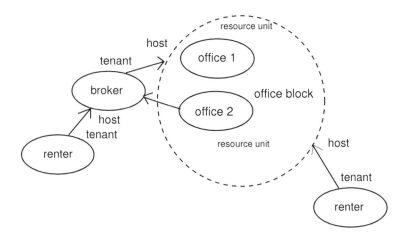

Figure 13.5: Identifying tenants and hosts correctly requires us to follow the tenancy law carefully. In each case, the arrows point towards to host resource sought by the tenant. (1) A renter may be a tenant of either an office block (providing multiple offices to multiple renters), (2) A renter may be a tenant of a broker (providing multiple client offices to multiple renters), (3) An office block or single office may be a tenant of a broker (offering multiple renters as a resource) to multiple office blocks or office.

Example 80. *In the OSI network model, the layers from L1-L7 form a tower of dependence, in which network resources (at the bottom) are shared out between different applications and users which are tenants of the basic service. These layers farther up the stack depend on the lower layers, hence the arrow of tenancy points down to the L1 physical layer. L2 is a tenant of L1, L3 is a tenant of L2 and so on.*

Network encryption is a tenant of L3, and computing applications are tenants of the encrypted stream.

When these layers are implemented as encapsulation, the tenancy increases into the core of the encapsulation, and the host is the outside part. This seems to be the opposite of the way we are taught to think about networking, from a software engineering perspective.

Lemma 9 (Causation is partially ordered by pre-requisite dependency). *Promises and intentions may be partially ordered by conditional dependencies, from the conditional promise law. This leads to a hierarchy of directional intent, for fixed semantics.*

We can distinguish tenancy from simple scaled agency by this partial ordering of tenants to hosts in the direction of a named resource. However, in most cases, the law of complementarity of promises allows us to transform one tenancy into the reverse

13.4. OCCUPANCY AND TENANCY OF SPACE

relationship interpreted as a different promise. In either case, the orientability of tenancy gives agents topological 'hair' which can be combed in a certain direction, as a vector field.

13.4.3 FORMS OF TENANCY

Let's look at some familiar exemplars to see how this general pattern is realized in different scenarios.

- **Club membership, or passenger with ticket**

    The issue of club membership is one where an agent associates itself as one of a group of typed agents: a vector promise directed to a specific host. The host offers the tenant a membership, and the tenant accepts the membership lease.

    $$C \;\to\; \text{membership fee} \tag{13.26}$$
    $$R \;\to\; \text{membership credentials} \tag{13.27}$$
    $$f(C,R) \;\to\; \text{benefits and services} \tag{13.28}$$

    Membership in a club is a label, i.e. a property of an agent. However, in the case that a separate agency validates this label as evidence of an association, we can view the members as guests of the hosting club. The condition C is typically some kind of subscription, the membership itself is promised with a badge or access credentials, and the additional services that accompany membership require showing of the credentials.

    If a club is exclusive, then the promise of $+R$ has finite valency, else it has infinite or unlimited valency.

- **Employment** An immediate corollary of membership is employment at an organization.

    $$C \;\to\; \text{work performed} \tag{13.29}$$
    $$R \;\to\; \text{employee status/badge} \tag{13.30}$$
    $$f(C,R) \;\to\; \text{benefits and wages} \tag{13.31}$$

    In this case, an employee is a tenant of the hosting company that pays for membership with his/her daily work. Tenancy is fulfilled by access or credentials (the company badge), and benefits include wages, lunch, travel costs, etc. Tenancy is always symbiotic, by nevertheless asymmetrical. The relative values of C and $f(C,R)$ are in the eyes of the beholders. When trading promises, what is valuable to one party is usually not valuable to the other, else they would not be motivated to trade.

- **Privileged access (territorial access)**

    A further corollary is the use of credentials to gain access to territories, e.g. foreign visas, password entry, identity cards, etc.

$$+C \to \text{identity credentials} \qquad (13.32)$$
$$-C \to \text{authentication/access control} \qquad (13.33)$$
$$R \to \text{access passport/visa} \qquad (13.34)$$
$$f(R) \to \text{territorial access/resources} \qquad (13.35)$$

- **Shared exclusive resource usage (multi-tenancy)**

    Now consider the case where we add a finite valency to a limited resource, as well as a condition of fair sharing. A fair sharing promise, up to a maximum valency of n, becomes an additional constraint on the host, of the form:

$$+R_j \#n \;\Big|\; \sum_i^n R_i \leq R, \qquad (13.36)$$

    for each qualifying tenant Tenant_j, paying its tenancy cost C_j.

    See figure 13.6

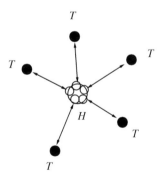

Figure 13.6: Tenancy of a singular resource by multiple agents. This is the same as membership, and containment. If multiple tenants occupy the same space, then the host effectively promises independent constituencies, so it has internal structure.

13.4. OCCUPANCY AND TENANCY OF SPACE

So the total promise set becomes:

$$\text{Host} \xrightarrow{+R_j \#n \,\big|\, C_j,\, \left(\sum_i^n R_i \leq R\right)} \text{Tenant}_j \quad (13.37)$$

$$\text{Tenant}_j \xrightarrow{-R_j \#m} \text{Host} \quad (13.38)$$

$$\text{Tenant}_j \xrightarrow{+C_j} \text{Host} \quad (13.39)$$

$$\text{Host} \xrightarrow{-C_j} \text{Tenant}_j \quad (13.40)$$

where $m < n$ by necessity due to the valencies. Services $f(C, R)$, based upon R might be subject to additional constraints, but they are also naturally limited by the constraint (13.36).

- **Representation by proxy** (spokesperson)

  In some cases a hosting agency's purpose is to be a proxy or representative for a client. This is the case for modelling agencies, writers' agents, sales representatives, public facing spokespersons, and even accountancy firms. Examples include 'Intel inside', goods on a shelf in the shop that represent their brands.

  In this case, the value added service to signing up is the representation of the tenant itself:

  $$+f(C, R) \to \text{Tenant representation} \quad (13.41)$$

  Representation or brokering for the tenant does not necessarily imply constraints on the tenant's autonomy (this depends on other promises). This is not exchange of the tenant, like sending a letter, or transporting a passenger. Notice, furthermore, that nothing promised here can prevent the tenant or host from acting as separate entities in other ways.

- **Catalysis** (special semantic environments)

  In a chemical process, some tenants need the help of a tailored environment to make a transition to a new state. A host plays the role of catalyst

  A pit-stop for tyre change, or a port/dock for loading and offloading, or repair of transport vessels.

  In the human realm, start-up labs and incubators are catalysts for companies and biological processes. The womb is a host for infant morphogenesis.

  In each case f(C,R)

Example 81. *Users are tenants of multi-user software, logging into walled communities with login credentials. Processes are tenants of operating systems. Operating systems are tenants of computer hardware. Computers are tenants of networks and datacentres.*

13.4.4 TENANCY AND CONDITIONAL PROMISES

It should already be apparent from the definition of tenancy, in section 13.4.1, that there is a likeness between the pre-condition for tenancy (denoted C) and the resource relationship (denoted R). From the conditional promise law, a conditional binding to provide service S takes the form

$$A_T \xrightarrow{+b} A_1, \left.\begin{array}{c} A_1 \xrightarrow{S|b} A_2 \\ A_1 \xrightarrow{-b} A_2 \end{array}\right\} \simeq A_1 \xrightarrow{S} A_2 \qquad (13.42)$$

Notice how the exchange of the condition has the same structure as the tenancy relationship. This is because both are examples of a generic client-server relationship, based on vector promises.

This can be formalized this further to show that a tenancy is really a conditional promise (see figure 13.7).

$$\begin{array}{ccc} H \xrightarrow{+R} A_? & \text{vs} & A \xrightarrow{-c} D \\ T \xrightarrow{-R} H & \text{vs} & D \xrightarrow{+c} \\ H \xrightarrow{+f(C,R)|-R} T & \text{vs} & A \xrightarrow{+b|c} A_? \end{array}$$

(13.43)

Thus the tenant is the assumed recipient of functional promises derived from the tenancy relationship, whereas in a general conditional promise this is unspecified.

Note, we shouldn't worry too much that the sign of the $+R$ maps to a $-c$, as the complementarity rule (see section 6.2.2) allows us to re-interpret the signs. For example, $+R$ could represent the active garbage collection of resources, while $-R$ represents quenching with resources. Similarly, $+R$ could represent employment, while $-R$ is work done to fulfill the employment moniker. In both these cases, the $+$ promise takes on the character of a receipt of service, often associated with $-$ promises.

The tenancy relationship is just an extended version of the basic client-server relationship, with the special focus on identity.

13.4.5 REMOTE TENANCY

If we consider the case in which tenancy is not between agents that are actually adjacent to one another, then the promises are delivered by proxy, in the sense of a delivery chain (see section 11.4, and figure 13.8).

When carried out via proxy, every adjacent node in a connective path through the adjacencies of the carrying spacetime becomes a possible point of failure or loss of

13.4. OCCUPANCY AND TENANCY OF SPACE

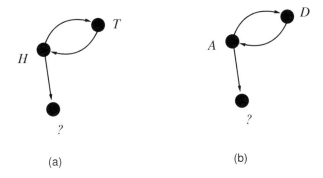

Figure 13.7: The likeness between tenancy and a conditional promise involving a third party shows that tenancy is not a unique phenomenon—it's a rather simple dependency.

integrity, and the cost of promising explicit integrity increases as the square of the number of agents along the path taken via adjacent agents.

13.4.6 ASYMMETRIC TENANCY

The semantics of tenancy are always asymmetrical, by definition. Adjacency is usually symmetric and mutual, at least when locations are equally weighted. However, if a superior location is next to an inferior location, according to some weighted importance ranking, then the symmetry is broken, e.g. pilot fish surrounding a whale, shops surrounding a mall. Unifying locations like malls, hubs, planets are natural host-roles to shops, spokes, and satellites, regardless of their relative size, because they connect agencies into an accessible nexus. Their 'size' may be thought of in terms of their network centrality[BBCEM10], for example, which gives them semantic importance.

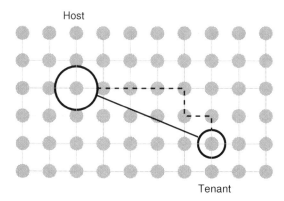

Figure 13.8: Promising tenancy virtually, over a substrate of truly adjacent intermediaries, often requires the distributive rule adds to the complications at the finer-grained scale, and scaling away these details requires implicit trust.

Lemma 10 (Adjacency is a form of tenancy, or tenancy is 'rich adjacency'). *By symmetrizing over the host-tenant promises, and directing unspecified promisees to mutual neighbours, we reproduce adjacency.*

$$H \xrightarrow{+R} (A_? = T) \quad , \quad T \xrightarrow{+R} (A_? = H) \tag{13.44}$$
$$T \xrightarrow{-R} H \quad , \quad H \xrightarrow{-R} T \tag{13.45}$$
$$H \xrightarrow{+f(C,R)|-R} T \quad , \quad T \xrightarrow{+f(C,R)|-R} H \tag{13.46}$$

which reduces to

$$H \xrightarrow{\pm R, f(C,R)} T \tag{13.47}$$
$$T \xrightarrow{\pm R, f(C,R)} H \tag{13.48}$$

Thus R plays the role of adjacency, and identifying $R \to \text{adj}$ and $f(C, R) \to \emptyset$, we see that adjacency is equivalent to mutual tenancy in its weakest form.

13.4.7 SCALING OF OCCUPANCY AND TENANCY

The ability to use space and time in a functional and operational way is the key to building organisms and organized processes. When we speak of scaling these semantic forms, we implicitly expect to preserve symmetries, asymmetries, and functional relationships,

13.4. OCCUPANCY AND TENANCY OF SPACE

while inflating the overall size of a semantic space by introducing more agents. Coarse-graining should then allow us to see the functional equivalence of the larger and the smaller system.

The asymmetry inherent in the ideas of occupancy and tenancy suggests that we are not generally going to see scale-free phenomena. What characterizes tenancy and occupancy is the retention of a differentiated cooperative relationship between agencies. Specific agents are bound together with intentional directionality. This contrasts with the idea of absorbing new agencies into a singular agency.

Example 82. *In a business partnership, or symbiosis, businesses or organisms retain their separate identities and work together for mutually beneficial returns. In a merger or acquisition, one company or organism subsumes the other, hoping to control it without worrying about explicit cooperation.*

The homogeneity of host-tenant semantics often play a role in the coordinated, functional usage of space. Long range order helps us to utilize space in a regular way. Without it, many aspects of space and time are simply opportunistic.

Example 83. *In a parking lot, the spaces need to be homogeneous in size else you might not be able to park your car in just any space. The same applies to the width or refrigerators, washing machines and kitchen appliances, block and sector sizes on disks.*

We need to account for both strong and weak couplings, homogeneity and inhomogeneity, to understand the wealth of possibilities in the world around us.

13.4.8 DISTRIBUTION OR DISPATCH OF PROMISES AT SUPERAGENT BOUNDARIES

So, what happens at boundaries when a promise is made to a superagency? What happens to the information? Let's try to answer the question dynamically first, since everything is dependent on what is dynamically possible.

When we make a new promise at scale M to a superagent, we need to understand what this means for the component subagencies at a finer-grained scale within it. Two possibilities present themselves:

- *Distribution/flooding (broadcast)*: Promise bindings made to a superagent are broadcast or diffused throughout the subagents that comprise it, spanning multiple agent locations, like the behaviour of a gas or fluid *flooding* into contact with an interface.

- *Direction/dispatch (switched)*: Promises are routed to a subset of subagents, or representative binding sites, in a solid state, making an exterior use-promise on the surface is responsible for accepting the promise. The routing can be direct from the promise to the interior subagent (if the superagent exposes its directory), or it can be made via a proxy routing agent inside the superagent (if it exposes only a gateway).

Why and how these possibilities should happen at all merits some further discussion. The details almost certainly depend on the scale and context. The generality of the questions (and the occurrence of examples in the natural and technological worlds) is what makes them most intriguing.

Example 84. *Consider an example of an extended superagent $\{a, b, c, d\}$ bound by some cooperative promises, which we neglect to mention here. These may occupy a space of similar extent $\{A, B, C, D\}$, as in figure 13.9. This scenario is a realization of many possible scenarios, e.g. a journey in many legs (plane, train, network routing), in which the promise of multi-tenant sharing of several sequential host resources forms a journey in which several hosts have to cooperate as a superagency (an inter-network cloud) (see figure 13.9). A traveller, (a tenant of the journey) has to be authorized for passage by each stage of the journey. This requires promises to authenticate credentials to be distributed throughout the path, and the collaboration of the hosting agencies in trusting the credentials.*

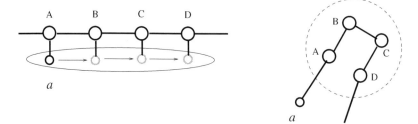

Figure 13.9: Path tenancy. The circled agents form a single superagent that occupies the corresponding space in the line above. The circles region is simply a superagent with exterior promises at its end-points. An agent binding to each site in transit can only do so at the scale of the subagents but the tenancy binding can be made as a distributive promise to the superagent.

Example 85. *Directing or dispatching promises, through a specialized agent, is like using a reception desk, service portal, in an office or hotel. Routing of information requires the underlying adjacency infrastructure to be able to direct messages to particular addresses.*

13.4. OCCUPANCY AND TENANCY OF SPACE

We may now state the two methods formally, for clarity:

Definition 63 (Distributive promise, at scale M (flooding)). *A promise made to a superagent A_s*

$$A \xrightarrow{+b} A_s \tag{13.49}$$

is assumed made to all agents within A_s:

$$A \xrightarrow{+b} A_i, \quad \forall A_i \in A_s. \tag{13.50}$$

The agents A_i voluntarily accept the promise, if they are suitable recipients, hence selecting by brute force rather than intentional labelling.

Example 86. *In information technology, flooding is used to make a 'bus architecture'. Ethernet and wireless transmission are examples.*

Definition 64 (Directed promise, at scale M (dispatch)). *A promise of type τ, made to the superagent, is assumed directed to a named subset of (one or more) members, on behalf of the entire superagent.*

$$A \xrightarrow{+b} A_i, \quad A_i \subset A_s. \tag{13.51}$$

The subset A_i voluntarily accept the promise, if they are suitable recipients, which we may assume is likely, given the intentional direction.

Example 87. *In information technology, dispatch to a directed address is used in queue managers, like load balancers, or memory and storage devices, to route data to a labelled destination.*

Stating these methods does not imply that they are possible in all cases. To understand whether diffusion of promise information is realizable we need to understand the small scale adjacency structure of spacetime, and its effect on promise scope.

13.4.9 ADJACENCY BETWEEN EXTERNAL AGENTS AND SUPERAGENTS

Can promises, made by an exterior agent, reach all the internal subagencies in a superagent, then be comprehended and accepted? A promise made to a superagent has to be transmitted along the network of underlying adjacencies.

Both dispatch and distribution approaches to dissemination and binding assume that promises can be made directly between the subagents of neighboring superagents. The communication needed to make and keep such promises depends greatly on the network substrate of adjacency made at the lowest spacetime level. So the question becomes one about how spacetime adjacency is wired (see figure 13.10).

Example 88. *To visualize promises made at coarse-grained scale, imagine a water authority that promises electricity to a town. Both these agencies are superagents composed of many subagents. Where (which agent) does the promise come from, and who receives it? What adjacency allows the promise to be transmitted?*

A generic promise made in the name of the company, depending on its legal department might make the promise. Every resident in the town is a potential recipient, as long as they can receive the information directly or indirectly, i.e. as long as they are in scope. The adjacency might be by postal communication and by water pipe.

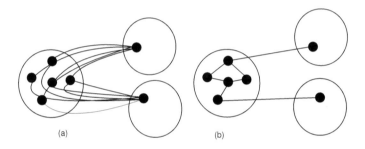

Figure 13.10: Regardless of whether promise diffusion has the semantics of flooding or directed dispatch, one is limited by the actual adjacencies that mediate communication in the space. In (a) agents are directly adjacent or 'patched' to all promisers of a particular type, and hence the promise diffuses naturally. In (b) adjacency is only to specific 'front desk' agents, who must then arrange adjacency virtually by proxy.

In figure 13.10a, the external agent promising to a superagent is directly adjacent to every subagent inside it. In figure 13.10b, the external agent only connects to a binding site. How these adjacencies come about, in practice, depends on the phase of the agents. There are two possibilities:

- Agents in a disordered gaseous state (no long range order), agents have no prior knowledge about one another without random walk meetings, and binding to one another. Thus discovery is a kind of Monte Carlo search[1], and communication is like broadcasting or flooding with messenger agents.

---
[1] This is somewhat like the way the immune system binds to cell sites.

- Agents in an ordered phase, can assign fixed coordinate locations which can be indexed and used to access agents by design. This requires a mapping in the index between promises and locations, and adjacency between the index and each agent inside the superagent boundary. Thus an index or directory service must act as a switch, routing promises to intended destination.

  This, in turn, can be done in two ways:

    - by exchange of agent contents at the boundary, and exterior lookup
    - by encapsulation of agent contents and routing with interior lookup

13.5 Human systems – 'soft topics'

At the human level, Promise Theory might seem simplistic—yet that simplicity exposes universal characteristics impartially in a way that social sciences tend to struggle with. It would be too much to expect Promise Theory to claim unique answers to every issue, but it's simple assumptions can nevertheless be a powerful tool for clarifying issues—and, in particular, it gives a clearer and less rhetorical meaning to the idea of 'social contract'. The challenge of any theory is not to lose expressibility in so doing, but rather to curtail or even deprive the ability of individual analysts who may wish to bend words to impose their own agenda.

13.5.1 Seeking cooperation and support of other agents

How do agents come together and interact to form collaborations and effective processes? This is obviously a complicated issue, and yet the simplistic notions of Promise Theory can still help to frame our thinking without getting bogged down with heuristics. Let's see what happens when we apply some of the principles.

We have two tools to use: promises and impositions, in both $\pm$ varieties. If we think about cooperative processes in companies, firms, institutions, clubs, or even countries, there is a mixture of command and control, and voluntary cooperation. Command and control suggests a kind of force or coercion involved in inducing collaboration. Even in democracies, there might be free voting but before that there is an imposition of candidates who want to force their ideas onto a population. They may claim to make promises which may or may not be kept. These may even be deceptions or lies. Our notions of freedom and imposition are clouded by context and emotional considerations.

In any system with competing objectives, the challenge of winning coherent support lies in condensing the available agents around a small number of alternatives. The larger the number of alternatives, the less stable a majority 'win' can be.

Example 89. *When voters promise to support democratic choices in an election, the counting of supporting votes may lead to problems of indecision and paralysis in different ways at different scales. The insistence on a majority becomes tricky:*

$$\exists i : voters_i \gg voters_j, \forall j \neq i, \tag{13.52}$$

Here, the meaning of $\gg$ is to be decided by some arbitrary margin. This may work well with few alternative choices and large numbers of voters who promise support for each choice, because the likelihood of a 'tie' (where no single agent satisfies the condition) is small, but the promise of a clear outcome becomes less probable as each alternative choice gets less support. If the number voters becomes too small, or the number of choices too great, one dilutes the relative strength of votes, so that it may not be possible to promise (13.52) for any choice i. Coalitions between groups, where two or more i join to form a single choice (to be determined by cooperation), may then be necessary in order for a superagent group to be able to promise a majority that can be accepted. Coalition involves a secondary level of cooperation, in which agents form superagents that now have the same problems on a smaller scale. In smaller elections one rarely sees democracy practiced, both because many voters would assess the opinions of only small numbers as arbitrary, and it could simply lead to paralysis and instability. The same issue occurs when a larger population votes for a representative, who in turn votes for them in a parliament or board of directors. Such intermediate agents vote with 'special powers' that promise to represent opinions of collections of voters—though the intermediate agent law (see section 7.2.2) shows that they may not be trustworthy. When voting in small numbers, the rules may accept a majority of one ($\gg$ becomes simply $>$), and the representative elections prefer odd numbers of voters to avoid tie-breaking during votes. The process of scaling is quite unstable to numbers, and there is a sense of misrepresentation—that the forms of democracy are followed without the promise of a majority having any significance. By careful engineering of categories, one could arrange for any result to win.

The balance of power between factions is therefore fragile when agents give their support voluntarily. This leads many to assume that the use of force is a necessary part of governance.

Force can take on several forms, from the use of an army to subdue opposition, to the insistence of compliance by *obligation*—an implicit threat. We know that, statistically, impositions of these kinds may be ineffective—not because of freewill or individual unwillingness, but simply because agents may be unable to comply with an intent that didn't originate with themselves. This suggests that there needs to be a balance between imposition and promise in any kind of governance process. That seems to be true at the lowest levels of physics (even if we don't understand why) but the nature of voluntary

13.5. HUMAN SYSTEMS – 'SOFT TOPICS'

and involuntary force has rarely been discussed impartially at the level of social systems. Instead, many authors have used Game Theory to imagine a conflict scenario.

13.5.2 Invitation, Attack, and Command

Let's take a simple promise theoretic view and consider an agent S of some size who wants to influence the cooperation of another agent R. There are some distinct types of communication from invitation to intrusion:

1. *Open Invitation*: A completely unconditional promise to either i) offer $+X$ to R, in the hope R might promise to accept:

$$S \xrightarrow{+X} R \tag{13.53}$$

or, conversely, ii) to accept $-X$ in the hope that R might promise to supply:

$$S \xrightarrow{-X} R. \tag{13.54}$$

In other words, an 'open invitation' is an opening in the form of a directed promise without coercive intent, e.g. promising to leave a door open in case someone wants to come in. The invitation does not suggest the correct or acceptable outcome, it only makes one available. We shall want to go further than this and express the meaning of a 'directed invitation' in section 13.5.3.

2. *Publish*: The generalization of an invitation is to publish to a wider audience, in the hope that agents may accept or even subscribe. This is a wider promise of availability.

$$S \xrightarrow{\pm X} * \tag{13.55}$$

For instance, the publishing of a magazine which interested parties may voluntarily choose to read.

3. *Command or targeted intrusion*: A directed imposition to accept or request:

$$S \xrightarrow{\pm X} \blacksquare R \tag{13.56}$$

Such an imposition might be an uninvited guest or an expectation of compliance with a directive. Steering a ship is an imposition in the sense that the ship promises to accept steering, but not a specific course. If a single course is accepted in advance, then the situation is reversed: the pilot is accepting a menu option promised freely.

4. *Broadcasting/Flooding*: The publishing of an imposition leads to a scaled intrusion, such as in broadcasting.

$$S \xrightarrow{\pm X} \blacksquare * \qquad (13.57)$$

For instance, even an *appeal* to voluntarily give money to charity is an imposition, because the message itself is forced into the recipients by media or by street collectors. It is a targeted imposition of a request or invitation. Marketing, advertising, and propaganda are examples.

From the complementarity between $+$ and $-$ promises, we know that there must also be an interpretation of every $+$ promise as a $-$ promise. For example:

1. *Stepping up*: A promise of $-X$ to accept an invitation:

$$S \xrightarrow{-X} R \qquad (13.58)$$

or a promise to provide $+X$ if an agent needs it

$$S \xrightarrow{+X} R. \qquad (13.59)$$

In other words, the complement of an invitation is a kind of generosity.

2. *Subscribing*: A broader promise to accept or provide to or from a variety of sources.

$$S \xrightarrow{-X} *. \qquad (13.60)$$

3. *Asking for a favour*: Targeting an agent to offer a promise that was not autonomously given.

$$S \xrightarrow{\mp X} \blacksquare R. \qquad (13.61)$$

4. *Obliging compliance*: Widespread inducement to promise certain behaviour, e.g. everyone must wear a helmet or fasten their seatbelts.

$$S \xrightarrow{\pm X} \blacksquare *. \qquad (13.62)$$

13.5.3 A MORE REFINED VIEW OF INVITATION VS ATTACK

In the foregoing section it emerged that invitations cannot be quite as simple as the distinction between promises and impositions, as invitations can themselves be imposed

13.5. HUMAN SYSTEMS – 'SOFT TOPICS'

$(\pm)$	$(\mp)$	INTERACTION
INVITATION	STEPPING UP	$S \xrightarrow{\pm X} R$
PUBLISH	SUBSCRIBE	$S \xrightarrow{\pm X} *$
COMMAND	ASK A FAVOUR	$S \xrightarrow{\pm X} \blacksquare\, R$
BROADCAST	COMPLIANCE	$S \xrightarrow{\pm X} \blacksquare\, *$

Table 13.1: Comparison of interaction types in coordinating autonomous agents, in a human context.

onto unsuspecting agents (e.g. a subpoena to appear in court), or promised (e.g. a special offer displayed in a magazine). There is thus a distinction between the delivery and the intention—an extra level of diplomacy that establishes a general basis for proposing a more specific outcome (see figure 13.11). It's helpful to separate these into two kinds of invitation:

- *Open invitation*—a non-exclusive invitation, which leaves an option open for matching agents to avail themselves of.

- *Directed invitation*—a formal announcement of the promise to accept interaction, made on an exclusive basis, i.e. targeted to a specific identifiable agent.

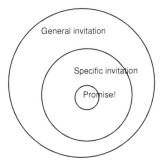

Figure 13.11: Invitation is a mechanism for homing in on a promise without a formal agreement, paving a way to agree by suggesting that a promise will be accepted.

Example 90 (Open invitation to shop). *A regular shop does not usually invite customers by mailing them a ticket to enter, or by promising to sell every good in the store separately (some catalogue services do this). Nor do shoppers accept or reject every offer. This would not be a scalable matter. Rather, some superagent scaling is involved. A shop* offers to be open for selling goods, and shoppers promise to accept these opening hours. This is the basic invitation framework for a promise dialogue. Customers then browse the goods, all of which follow the generic pattern: 'shop promises to sell if you pay a given price', and the shopper equally promises 'I promise to pay a given price if you sell me the goods'. This deadlock is broken by the customer imposing a choice and paying up front.

Example 91 (My charge is $+e$). *In physics, particles of different kinds advertise different kinds of charge (electric charge, colour charge, and so on). These play the roles of open invitations for other agents with charges of the same type to interact with them. The idea of an electron sending an RSVP to its neighbour is somewhat ridiculous, though messenger particles do exist for the charge interactions. The semantics of interactions are not directed by identity—electrons and quarks are indistinguishable, so their promises are non-exclusive.*

Example 92 (Invitation and deadlock in commerce). *If a shop leaves its door open for all to enter and buy goods, this is an invitation. The shop also promises each good for a price. Shoppers may promise to pay the price for a good if available. The deadlock between: 'you can have the good if you give me the money' and 'you can have the money if you give me the good' is broken by the imposition of voluntary payment by customers. In service interactions, service providers usually have to be the ones to break the deadlock by performing the service first. The service is less tangible than the good so its reality may need to be established before trust is built.*

Example 93 (Service negotiation). *Expanding on the simple view in the previous example, there can be several layers to a service contractor interaction, with layers of promises playing the role of invitations converging on a desired outcome.*

A contractor may advertise the availability of its service, which acts as an open invitation to hire its service. Acceptance of that leads to a problem discussion with a potential client, perhaps a Purchase Order (PO) for the solution, followed by an invoice for the purchase order (whether the solution is achieved or not) and payment of the invoice. There are layers of invitation here. Let's label the service provider or contractor by S, and the client as C:

$$S \xrightarrow{+skill, availability} C \quad (open\ invitation\ to\ hire) \tag{13.63}$$
$$C \xrightarrow{-availability} S \quad (accept) \tag{13.64}$$
$$C \xrightarrow{+problem} S \quad (explain\ need) \tag{13.65}$$
$$S \xrightarrow{-problem} C \quad (listen) \tag{13.66}$$
$$S \xrightarrow{+solution|PO} C \quad (listen) \tag{13.67}$$

13.5. HUMAN SYSTEMS – 'SOFT TOPICS'

The contractor takes some risk by discussing the problem up front, but withholds the solution, instead inviting the client to make a purchase order, and promising the solution conditionally on the PO. This shifts risk over to the client since the purchase order will be invoiced unconditionally on a solution.

$$S \xrightarrow{-PO|problem} C \quad \text{(invitation to commit)} \tag{13.68}$$

$$C \xrightarrow{+PO} C \quad \text{(accept and commit)} \tag{13.69}$$

$$C \xrightarrow{-solution} S \quad \text{(accept and trust promise of solution)} \tag{13.70}$$

An invoice is promised by implication of the PO:

$$S \xrightarrow{+invoice|PO} C \quad \text{(invite client to pay for the PO)} \tag{13.71}$$

$$C \xrightarrow{-invoice} S \quad \text{(accept)} \tag{13.72}$$

$$C \xrightarrow{+pay|invoice} S \quad \text{(accept and trust promise of solution)} \tag{13.73}$$

$$S \xrightarrow{-pay} C \quad \text{(accept payment)} \tag{13.74}$$

Note that payment is not conditional on the solution outcome in this model—only on the promise of a solution.

The example above illustrates the rich dynamics that can be involved in interactions on the complex level of semantic systems. This is not a unique feature of human systems. Clearly, the role of an invitation, in the human sense, feels like much more than simply leaving the door open. It may consist of many layers, and a gradual convergence of wooing the parties to cooperate. This is also seen by animal species in the wild during mating! It's an auxiliary communication—an announcement of intent to grab attention, which may itself be imposed or promised passively. By announcing an intent to invite one may even increase the perceived value of potential acceptance.

Example 94 (Don't drop litter!). *Advertising and propaganda that aim to teach people not to drop litter may be framed as invitations. These may be imposed upon them, for the public good, or targeted at individuals, depending on the scale of the targeting. It could be presented as 'you'd better do as your told or else' or as 'we invite you to be a better person'. Both approaches have been used in different countries*[59].

Invitations are sometimes handed out as promises and sometimes as impositions. The essence of an invitation is that one first tries to establish a promise acceptance at a general level to pave the way for the specific outcome. There is a progression of constraints that home in on the final targeted interaction, edging forwards 'is it okay if we do this? Good! Now what about this?' Invitation is a convergent surrogate for a process of agreement, in which no formal agreement is ever made. So how shall we decide whether the imposition of a decision for change by a manager, leader, or exterior entity of any kind is made by diplomatic invitation or by attack? What is the difference between these two? How are they defined and how can negative behaviours be avoided?

To reveal this in more detail, we may first nail down some underlying definitions pragmatically, without getting too deep into nuances. The following definition helps to show how a directed invitation works as a proxy:

Definition 65 (Directed Invitation to do X). *A prior promise or imposition of $I(-X)$, conditional on X is an invitation to accept a promise of $+X$:*

$$\text{Inviter} \xrightarrow{-X} \text{Invitee} \quad (13.75)$$

$$\text{Inviter} \xrightarrow{+I(\mathbf{def}(-X))|-X} \blacksquare \text{ Invitee.} \quad (13.76)$$

may be considered an invitation for the invitee to promise:

$$\text{Invitee} \xrightarrow{+X|I} A_?, \quad (13.77)$$

assuming that the invitee accepts:

$$\text{Invitee} \xrightarrow{-I(X)} A_? \quad (13.78)$$

$$\text{Invitee} \xrightarrow{--X} A_?. \quad (13.79)$$

The latter may be considered incorporated by the promise of $+X$.

We can read these as follows: (13.75) says: there exists an X that the inviter will accept unconditionally, and (13.76) says: given that acceptance of X, The inviter issues an invitation $+I$ for invitee to accept by promising $-I(X)$, and ultimately $+X$. Notice that it is the acceptance $(-X)$ of the promise, by the Inviter, which is unconditional, and the invitation merely encapsulates knowledge $\mathbf{def}(X)$ about its proposal, without assuming the Invitee would know about it in advance. The promise of $-X$ alone may not be sufficient information to imply a proposal to match it with $+X$. The role of an explicit invitation $I(X)$ is to signal a clear promise, rather than merely laying out a path to completion. Note that, by the rules of conditional promises, the two promises (13.75) and (13.76) are identical to an unconditional promise of $+I(\mathbf{def}(-X))$, so the promising of an unconditional invitation may also be conversely reinterpreted as an explicit acceptance of a raw promise of $+X$.

For example, suppose X is 'your attendance of an event'. The inviter promises to accept the attendance unconditionally of the invitee. This offer of open acceptance may be ambiguous, so the intent to invite is emphasized with an explicit promise of $I(\mathbf{def}(-X))$ which advertises the content of the promise $-X$, so the invitee knows

13.5. HUMAN SYSTEMS – 'SOFT TOPICS'

what it's getting itself into. This is what we call a directed invitation. It's a positive offer of the complementary promise! Note that each agent is the originator of its own promise. The response (hopefully a full $+X$) might only be a partial overlap with the $-X$ offered. Nothing is imposed by the other party.

The opposite of an invitation is an attack. This is the imposition of an offer $+X$ before establishing grounds for general acceptance of whatever X might be.

Definition 66 (Attack on agent R). *Any imposition on a recipient agent R by a source agent S:*

$$S \xrightarrow{+X} R, \tag{13.80}$$

for which there is no prior invitation or acceptance by R:

$$R \not\xrightarrow{-X} S. \tag{13.81}$$

The definition of 'attack' in Promise Theory is thus an attempt to induce cooperation without acceptance or invitation. We can't claim that invitations are promises and attacks are impositions unequivocally, because often there are mixtures of the two in communication. Communication might be voluntary, but its content imposes, or vice versa.

Example 95 (Covert advertising—hidden impositions). *A magazine is published and distributed for people to take freely. The reader promises to accept unknown article content, but does not promise to accept advertising. So there is a hidden imposition in the promise.*

Example 96 (Overt advertising—imposed promises). *Flyers are posted through your letter-box, or messages are pushed into a social media channel. The messages promise a free sample, but the users did not promise to accept messages of this kind.*

Example 97 (Public education). *We might ask: at what point does a public service become an imposition? The promise theoretic answer seems to be: when the service was not initially sought by its intended users. The distinction between 'spam' or propaganda, news, advertising, etc, is much less universal than we tend to think. The perceptions change by scale, by circumstance, and by perspective. The role of invitation thus seems to be to home in on a specific*

Interactions between agents, which iterate back-and-forth in a particular sequence, are the chief mechanism by which agents accumulate a foundation of promises on which trust is built. As with any house of cards, a stack of dependent promises is fragile to the order of proposals. We see this just from the two definitions above. The attempt to exploit the fragility of such dependency is how one disrupts, perhaps with an intent to attack. We should therefore expect to have to spend some time uncovering the consequences of strong ordering constraints in dependency relationships.

Example 98 (Attack or lead by example?). *If an imposition is made* ad hoc *before an invitation has been given, it may be assessed as an attack, with destructive consequences. On the other hand, if the imposition was expected, it may be considered an initiative.*

Attack	Invitation
$S \xrightarrow{+\text{initiative}} \blacksquare R$	$S \xrightarrow{-\text{listen/talk}} R$
	$R \xrightarrow{+\text{listen/talk}} S$
$R \xrightarrow{?} S$	$S \xrightarrow{+\text{initiative}\mid\text{listen/talk}} R$
	$R \xrightarrow{-\text{initiative}\mid\text{listen/talk}} S$

In the table above we see the contrasting approaches between a blunt attack on the autonomy of the receiver R, and the route of diplomacy on the right. Through invitation, the sender begins by promising (up front) it's willingness to listen and talk by promising its acceptance (-) of any promise about talking and listening the receiver might offer. The receiver has now been invited to cooperate, without imposition. If it quenches that invitation by promising to supply dialogue (+), the sender can then promise initiatives, building conditionally on the prior acceptance of the invitation, because by the rules of Promise Theory, an initial offer of a (-) promise (an invitation) is accepted with another (-), making a (--) which is equivalent to (+)[60]. This is now a promise rather than an imposition, since the prior invitation was sought and the proposal was expected. The receiver may then accept the promise of the initiative, or it might decline it at its option. It has only promised to listen, not to accept, so there is no attack. We see how diplomacy is the use of promises rather than impositions to open dialogue between agents. The perception of a violation of autonomy is an important assessment in cooperative mechanics, and clearly plays a role in human assessments of trustworthiness.

Example 99 (Layers of invitation). *Diplomacy is a process of homing in on a sensitive area by seeking invitations back and forth.*

'Would it be a terrible imposition of me to make a suggestion?'
'I invite your suggestion, based on your thinly veiled but humble meta-invitation.'
'Then I propose that ....'
'I accept your proposition, but reject part of its content....which needs revision.'
'I accept your proposal for revision of my proposal....'
And so on...

13.5.4 THE INTENT TO DISRUPT OR SABOTAGE

We can't properly discuss systems or bodies of systems without acknowledging that they exist within a larger environment of agents, some of which may interfere (intentionally or otherwise) with the interior promises and its intended behaviours. Since the term *interference* has a special meaning in physics, I'll choose the term *disruption* for this kind of conflicting interaction.

Definition 67 (Disruption of intent). *A promise, imposition, or the withdrawal of either, made by an agent D (called the disrupter), which renders a promise from S to R to be 'not kept', according to the assessment of some observer O:*

$$\alpha_O \left(S \xrightarrow{+b} R \right) \to \text{not kept.} \tag{13.82}$$

Looking at the assessment in the box above, the Principle of Autonomy in Promise Theory (that an agent cannot make a promise on behalf of another) implies that disruption can only be classified into three forms:

1. *Disruption of the First Kind*: the agent S intentionally fails to keep its promise to R. Then $D = S$. This is a kind of self-sabotage, i.e. a deception to R.

2. *Disruption of the Second Kind*: the agent S withdraws or fails to promise observability to the agent O assessing the outcome. $D = S$. This is a kind of subterfuge, i.e. a deception to O.

3. *Disruption of the Third Kind*: suppose that the promise body is conditional on a promise of Δ a third party T, i.e.

$$S \xrightarrow{b \equiv X | \Delta} R \tag{13.83}$$
$$T \xrightarrow{+\Delta} S \tag{13.84}$$
$$S \xrightarrow{-\Delta} T \tag{13.85}$$

where Δ becomes a condition for S to keep its promise of X, which fails to keep its promise, or attempts to impose a counter-intent. Then $D = T$. This is an external 'man in the middle attack' by D. In this case, we have:

$$D \xrightarrow{+\Delta} S, R, O, \ldots \tag{13.86}$$
$$S \xrightarrow{+X|\Delta} R \tag{13.87}$$
$$S \xrightarrow{-\Delta} D \tag{13.88}$$

The promise Δ may be withdrawn by D if positive, or negated. Any change to Δ applies influence or disruption to the intended outcome. Equally, S may fail to accept this promise (13.85) (through no fault of T), but this is covered by disruptions of the First Kind. Extreme cases of disruptions of the Third Kind are:

$$\Delta \to \chi, \qquad D \xrightarrow{\emptyset...\neg\chi} \blacksquare\, S, \quad X|\Delta \to \emptyset \qquad (13.89)$$

$$\Delta \to \neg\chi, \qquad D \xrightarrow{\chi} \blacksquare\, S, \quad X|\Delta \to \emptyset. \qquad (13.90)$$

The promise model is quite helpful in clarifying the roles of the agents in these interactions, without need for speculation. Disruptions can be caused at the source, by the receiver, or by intermediate agents. The duality rule, citing the equivalence of juxtaposing $\pm$ promises in a binding, implies that the same argument applies when the roles of sender (+) and receiver (-) are exchanged. Most of us tend to ignore the essential promise to accept:

$$R \xrightarrow{-b} S, \qquad (13.91)$$

yet the receiver can disrupt a promise-binding by refusal to cooperate, as well as refuse to allow an assessor or auditor of the promise to access to its behaviour. The straightforward outcome of framing disruption in that we should be on the lookout for hidden dependencies:

Lemma 11 (Third party disruption implies dependency). *Only a conditional promise can be disrupted by a third party, implying a dependency in the original promise body.*

13.5.5 Accusations, the imposition of judgement, and taking offense

An issue related to the foregoing examples is the matter of the taking offense at remarks made by agents to others[61]. This follows immediately from the dual status of promises to give (+) and to accept (-). It is every agent's autonomous nature to form its own judgements, as well as to express them. However, as already established above, the imposition of one's judgement may be perceived as an attack, is made uninvited. This is clear when the framing of the imposition is itself unambiguous:

$$S \xrightarrow{+\text{You are a jerk}} \blacksquare\, R. \qquad (13.92)$$

This is an attack, unless permission was granted by R

$$R \xrightarrow{-\text{You are a jerk}} S. \qquad (13.93)$$

S's promises or impositions are also subject to intentional misrepresentation by R in the overlap between what is promises and what is understood. Suppose, S issues a declaration:

$$S \xrightarrow{+G \text{ is a nice type}} A_?. \qquad (13.94)$$

13.6. RESPONSIBILITY

This is not an attack, since it is promised freely to an unspecified recipient, without any attempt to oblige a specific recipient R. Nonetheless, R may still use this to attack S however, by interpreting 'type' as a deliberately condescending term for a 'person'. This is a consequence of self-duals in Promise Theory that interpretation can be an attack on an agent's utterance simply by imposing its own (different) interpretation. If the sender intended the phase as a playful and complimentary expression for a ninety year old G, then the conflict becomes an imposition. Similarly, the receiver might object to the meaning of 'nice'. 'How dare you call me nice?' Too much, too little? The art of taking offence or offense is a specialized form of attack by reverse proxy.

Overt Attack	Reverse Hijack
$S \xrightarrow{+X} \blacksquare R$ $R \xcancel{\xrightarrow{X}} S$	$S \xrightarrow{+X} R$ $R \xrightarrow{-X} \blacksquare S$
Intent imposed without invitation	Interpretation imposed without intent

13.6 Responsibility

Responsibility is an emotionally charged term. It is associated with liability, blame, and reprimand. There is a tendency to want to assign 'blame' to a person when faults lead to loss. That's a punitive interpretation of responsibility, and one that does not often make sense, unless willful malice or negligence were demonstrably at work. We need a more rational understanding of responsibility, based on the science of causation, to actually stand a chance of making a difference. If we can step back from the idea of attributing blame, there is something to be learned from asking the question: which agency or agencies are in a position to be *able* to keep a promise, and hence *could* be considered responsible?

13.6.1 Subjectivity in assessments

The assessment that a promise has not been kept is a subjective one: different agents observing the system might assess it differently. Their assessments can depend on context or circumstances; so how can we easily attribute a unique source to the perceived failure? This is the challenge of a distributed system with multiple stakeholders.

Example 100. *In a restaurant, a meal is ordered from the menu. One person enjoys the meal, the other doesn't. The latter (dependent agent) assesses the meal to not be what was promised. The former (fault tolerant agent) assesses the meal as acceptable.*

- The agent that rejects the meal might be considered discerning of quality, but goes hungry and cannot work.

- The agent that accepts the meal eats and continues its work.

Can a cause be attributed to the waiter, the head chef, the sous chef, the butcher?

Whether or not a promise has been kept depends on the kind of promise binding. As we know about promise bindings, the receiver or promisee has to make its own promise to accept what is offered; thus it shares responsibility in outcomes. Indeed, refusing to accept what is offered is the ultimate control decision of autonomous agents.

Example 101. *A doctor promises a patient: if you take these pills you will be cured. If the patient does not keep a promise to take them, then it will not be cured, and thus the responsibility for being cured lies ultimately with the patient, not the doctor.*

13.6.2 THE ROLE OF CONDITIONAL PROMISES IN POINTING TO RESPONSIBILITY

In promise theory, we track provenance or causation with *conditional promises*. Each promise is the responsibility of the agent who makes the promise (the promiser). From the conditional promise law, an agent making a conditional promise has not made a promise at all unless it also promises to acquire the thing its promise is conditioned on.

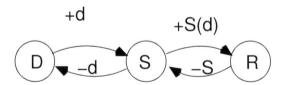

Figure 13.12: A dependency chain from upstream dependency provider D to a server S relying on the dependency, to a downstream recipient R.

Consider the scenario in figure 13.12

$$D \xrightarrow{+d} S \tag{13.95}$$

$$S \xrightarrow{-d} D \tag{13.96}$$

$$S \xrightarrow{+S(d)} R \tag{13.97}$$

$$R \xrightarrow{-d} S \tag{13.98}$$

13.6. RESPONSIBILITY

where

$$S \xrightarrow{+S(d)} R \equiv \begin{cases} S \xrightarrow{+S|d} R \\ S \xrightarrow{-d} R \end{cases} \quad (13.99)$$

This system is fragile because the recipient has only a single choice. It has a single point of failure. The recipient could seek out redundant alternatives to provide the service S (as in figure 13.13). What happens beyond the horizon of the next agent in the chain of promise relationships is beyond the control of the recipient, and is thus beyond the limit any possible responsibility. Now consider the same scenario with redundant alternatives

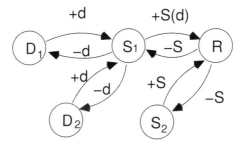

Figure 13.13: A redundant conditional promise chain, showing service delivery based on a dependency.

along the chain (see figure 13.13)

$$\{D_1, D_2\} \xrightarrow{+d} S_1 \quad (13.100)$$
$$S_1 \xrightarrow{-d} \{D_1, D_2\} \quad (13.101)$$
$$S_1 \xrightarrow{+S|d} R \quad (13.102)$$
$$S_1 \xrightarrow{-d} R \quad (13.103)$$
$$R \xrightarrow{-S(d)} S_1 \quad (13.104)$$
$$R \xrightarrow{-S} S_2 \quad (13.105)$$
$$S_2 \xrightarrow{+S} R \quad (13.106)$$
$$R \xrightarrow{-d} \{S_1, S_2\} \quad (13.107)$$

In this second scenario, both the server S_1 the recipient can choose from two providers of the promises they are trying to use. For the final recipient R, the fact that the promise from S_1 has a dependency is irrelevant, as there is nothing it can do about that except to acquire a second provider who may or may not have a dependency too. The only security the recipient R has is to have a choice of providers. No matter how hard the providers S_1 and S_2 try to keep their promises of service, unforeseen circumstances may prevent them from doing so. Indeed R may itself be negligent receiving their services.

This suggests that, while responsibility for keeping a promise lies with each source agent, only the final recipient (the location of the desired outcome) can be considered responsible for securing a successful promise outcome.

13.6.3 DOWNSTREAM PRINCIPLE

Locality thus gives a surprisingly simple and consistent interpretation of responsibility. The recipient of a promise carries the ultimate burden of assuring the outcome. In a chain of promises, dependencies are upstream (the source of the flow of influence) and the benefactors are downstream. Based on the consistency of responsibility described above, we can make the following straightforward observation. The assurance of the final promise outcome follows a 'downstream principal' that the most downstream agent has both access and opportunity to correct or absorb faults, and hence the greatest causal responsibility for an assessment of a promise not being kept. In other words, the greater the distance from the point of promise-making, the less causal responsibility an agent has in contributing to the outcome.

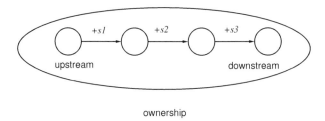

Figure 13.14: Responsibility for success in a chain flows downstream. The final responsibility is to use what is offered. An owner may formally accept responsibility for the whole chain, as a superagent, but at the microscopic level there is no other agent that can keep that promise. According to promise theory, the owner of the workflow should be the last interior agent to accept or 'sign off' on the work.

It's important to understand that the downstream principle is not a moral assessment, it is a purely pragmatic observation about cause and effect. However, it is interesting that it is in opposition to what is conventionally assumed about faults in hierarchies, as well

13.6. RESPONSIBILITY

as in Root Cause Analysis. The explanation for this apparent contradiction can be found in the bi-directionality of promise bindings required for propagation of influence. The traditional assumption has been that influence is always imposed and that impositions always succeed, the latter being false.

Can we assert, then, that an agent who fails to use a promised service in order to keep its own promise is more responsible than the failure of the agent to provide the service? The user of the service could, in principle, seek a redundant alternative for such cases. But what if no alternative is available? If a promise is made conditionally, the agent (promiser) advertises a delegation of its responsibility to keep the outcome, by adding conditions. The promiser of each link in a chain of cooperation is responsible only for its own promises, i.e. the point at which it can effect change of behaviour or intent. This tells us that delocalization means divesting responsibility to others. This is the power of agent autonomy. Thus, in promise theory, responsibility is connected to locality. How does this compare to conventional component reliability theory?

- In the black box tradition of reliability theory, there is insufficient information to be able to attribute provenance for observed failures, so one can only attribute failure to a component itself. The failure of a component must be considered a random event, and the lack of information means that it makes no sense to assign moral blame.

- In Promise Theory, the assumption is that an agent that makes a promise is the only agency responsible for keeping the promise. It might be possible to argue moral or causal blame to the specifically documented promise. This is in keeping with the tradition of legal responsibility.

- By the conditional promise law, a promise that is conditional on another promise being kept (either by the same agent or by a third party) is not a promise, unless the other promise is made by the same agent. This clarifies and documents diminished responsibility.

We can now attempt a limited but tenable definition of responsibility:

> **Definition 68** (Responsibility for keeping a promise). *Responsibility has two main interpretations:*
>
> - Causal responsibility: *When an agent relies on a dependency promise in order to keep its own conditional promise, causal responsibility refers to the agent's freedom to obtain a promised outcome by its own autonomous choice of interaction, especially in the presence of redundant alternatives.*
>
> - Moral responsibility *(culpability) is a human assessment, about whether agent outcomes stem from good or for bad intent, hence it cannot be formalized except as a norm or in law. This is not a systemic issue, only a subjective assessment.*

An agent cannot be solely responsible for keeping a conditional promise. Could an agent that relies on a promise from another be culpable for a failure to find an alternative if the dependency fails to keep a promise upon which it relies? This might be considered a case of negligence, unless the agent's hands are tied by other considerations. The network of promises in which a agent finds itself ultimately determine these freedoms.

13.6.4 Assuming responsibility

Sometimes we hear leaders 'assuming responsibility' for their organizations. Here we are at risk of confusing responsibility with liability. An agent can 'assume responsibility' by making a symbolic promise. However this might not be a promise it can actually keep.

Causation implies that there are limits to what such a agent can really do. Promising to assume responsibility might be an unrealistic and symbolic promise, one that cannot be kept, since a single agent is potentially promising to be the single point of delivery for an entire organizational superagent. At best such a promise could only be conditional on promises made by the other agents in the organization, and thus its reliability would only be weak. By the same token, judging or holding an agent responsible is an imposition that may not be reasonably accepted.

We can make a rough prediction, based on the foregoing discussion: the only case in which it is not redundant to impose blame on another agent S for an outcome, is when the agent R (in the role of a downstream user (-) or consumer of a service) has not promised sufficient redundant contingencies for its own role in the responsibility for the outcome. Blame is therefore used as a way of deflecting attention from missing for the lack of keeping one's own promises.

13.7 RIGHTS, PERMISSION, AND PRIVILEGES

Rights, permission, and privilege are all concepts related to the giving of promises. The concept of 'rights' is an important issue in human systems: e.g. human rights, right of way, right of settlement, right of refusal, access rights, etc). In society, rights are granted to access medical care, to enter a country with a passport, and so on. When these rights are granted discriminately, they may also be considered privileges. In human-computer systems, for instance, access rights are granted to only certain users to see certain files or use certain programs. These form a part of what we usually call 'security' or 'privilege'. There are subtle distinctions between the semantics of security and privilege: the former is about safety and protection, the latter is about rank and hierarchy—yet the two reduce, for all intents and purposes, to the same basic issues of promises. The offer of certain promises (or not) amounts to the management of boundaries and realms of access.

The terminology of rights is common in politics, often used with a deliberate intent to fire up emotions, to impose and to attack the agents—such as governments—who make promises to provide services, and who are interpreted as having greater privilege because they possess capabilities that may not be promised to all agents that seek to use them. Demanding rights is an imposition, thus we expect the seeking of rights and impositions to be related too. The widespread abuse of the terminology of 'rights' stems largely from the historical origin of moral righteousness, and the confusion between *permissions* and *capabilities*, which in turn stems from a tradition of deontic or obligatory thinking. Promise Theory makes simple and impartial sense of these matters.

13.7.1 RIGHTS DEFINED

It follows from the assumption of agent autonomy that agents can only promise capabilities they possess autonomously, and that they may not promise capabilities of other agents on their behalf[62] (see figure 13.15). This has the immediate implication that

Figure 13.15: Rights are simply the status of promises about the capabilities offered by some agent. The absence of a promise X is not the same as a promise of its absence ($\neg X$).

'rights' cannot be innate and universal[63]. They are offered as promises by an agent about matters it can keep itself. Rights, in other words, are not something that are innate or fundamental properties of a system, but rather they are promises that may be offered by capable providers, perhaps as part of a larger social contract.

Definition 69 (The right or permission to X). *A promise, by a provider agent S to offer access to a resource or behaviour X:*

$$S \xrightarrow{+X} A_?. \qquad (13.108)$$

This promised 'right' is not an obligation to accept, so the pre-requisite to a binding is captured precisely by the concept of a (+) promise. We may also speak of permission to behave in a certain manner. Permission is a promise of access. The semantics of permission are only subtly different from those of a 'right'. When seeking permission, one assumes that the promise $-X$ to use X exists prior to the matching offer of $+X$ which grants access to it. In all other respects permission and right are equivalent. Such rights or permissions may be granted specifically to a named individual, or may be granted to any unspecified agents. If the provider discriminates between its promisees, then an observer might assess different levels of privilege for those agents.

Definition 70 (The privilege of X). *An assessment or promise of status attributed to one or more agents, based on its access to private resources. If an agent R is promised $+X$, i.e. is granted the right to avail itself of X:*

$$S \xrightarrow{+X} R, \qquad (13.109)$$

In this case, we may say that R is granted X privileges. The promise of access acts as a label of privilege, which is used as a token of status. This could be made explicit as a promise of rank:

$$R \xrightarrow{+rank(X) \mid X} *. \qquad (13.110)$$

i.e. given a promise of X, R may claim to be of rank X.

Again, the recipient granted or promised privileges is under no obligation to match the (+) promise with an acceptance (-). So the expression 'rank hath its privileges' is in fact the wrong way around: the promise of privileges are the foundation that define rank. Rights are therefore related to the notion of assisted promises (see section 6.2), i.e. a promise enabled by the granting of access to another promise it depends on.

13.7. RIGHTS, PERMISSION, AND PRIVILEGES

13.7.2 SEEKING RIGHTS AND PERMISSIONS

Agents may not be granted the rights and privileges they need to make their own promises *a priori*. If they are unconstrained by other promises to search for providers of what they seek, they apply their downstream responsibility (see section 13.6.3) to secure their desired outcome. Agents without the freedom to search freely may want to demand additional rights from the agents they are clients of. This assumes that they have an expectation about what services can and should be promised to them by another agent (e.g. a government, an employer, etc). The basis for this belief could originate in any number of ways, and it could simply be a fortuitous search for a match, as in evolutionary processes.

Example 102 (I know my rights!). *Agents may imagine or be programmed to respond to possibilities for which they have no matching promise.*

- *I have an electric charge and I am looking for a field to drive it.*
- *I have a hammer, and I am looking for a nail to complete it.*
- *I want the right to settle in Canada, and I think you should let me.*
- *I am lonely, and I demand a soul-mate.*

The rule of autonomy implies that they cannot demand compliance with these missing complements. They can only stand ready to accept promises offered to them from different sources.

Example 103 (Demanding one's rights). *The attempt to demand rights is an imposition to use a resource*

$$User \xrightarrow{-X} \blacksquare Provider. \qquad (13.111)$$

The strategy of demanding is likely to be ineffective since the provider may be unable or unwilling to comply. A strategy of invitation or search for alternatives is more likely to succeed.

A provider S only has the capability to deny another agent R access to X if it has interior private access to X, and R relies on other promises by S for its survival. We also say that the agent has the authority to grant permission (see section 13.8), meaning that it is the custodian of that privilege.

Example 104 (The right to free speech). *Any agent R has an a priori* ability *to promise speech of any kind, unconditionally and autonomously. The question of a* right *to speech only arises when some agent(s) S, with which R interacts, imposes (threatens) sanctions* to try to prevent an agent from exercising that ability. The implication is that R has already accepted a promise made conditionally on not exercising its capability for free speech from this other agent S, and this is being used as leverage—e.g. a promise that might be withdrawn, or a new promise which might harm R.

$$S \xrightarrow{+food|\neg speech} R \qquad (13.112)$$

The ability to speak freely thus precedes any 'right' granted concerning speech. The point is rather than the subject S has promised not to speak freely in order to receive promised benefits from S. In principle, R has the downstream responsibility to seek a replacement source for its needs if they are threatened. In common speech we commonly muddle this ability with language such as 'the fundamental right to free speech'. This is just imprecise language. The need for permission to exercise a capability one already has is an idea that can only arise in a cooperative framework in which voluntary abstention is practiced.

Rights—especially human rights—are something we often argue for, assuming that we deserve them. In Promise Theory, such a moral demand for rights is contrary to the autonomy of agents.

13.8 Authority, power, and delegation

The Oxford English Dictionary defines authority to be 'the power or right to influence others or act in a specified way'. Promise Theory tells us that there are two disjoint issues in that statement: respectively the 'power' and the 'right' to propagate influence—influence by command or by invitation. These two branches correspond loosely to the ad hoc *imposition* of influence, versus the *promise* of permission to wield influence, respectively.

Conventionally, we are more conditioned to think of the concept of authority through power. Most nations' legal stability is based on the threat of being able to overwhelm deviations from lawful behaviour by some kind of force—only later do such behaviours become norms and habits that require only simple maintenance. As a result, we perceive governments, bosses, and leaders in a historically authoritarian light (even the word has even come to take on a pejorative meaning, versus 'authoritative' which is more positive). In other usage, an authority is a source of singular expertise on a particular subject.

The unifying concept of authority is ultimately the appointment of 'trusted' agents[64]. An authoritative agent is one that promises to calibrate the definition of a kind of another promise on a particular subject. We define authority as follows:

13.8. AUTHORITY, POWER, AND DELEGATION

> **Definition 71** (Authority for X). *An agent which is the source of a promise with body X, and whose information is accepted by other agents, which all assess it to be authoritative on the matter of X. In other words, such an agent is a trusted party, according to the subordinate agents.*

The circularity of the definition reveals that authority is not an absolute property; it is only a self-consistent appointment and assessment, made on trust. The source is said to be authoritative, because it promises to be the final word on what is correct, and the subordinates accept that promise. That does not make it unique, as other authorities may have a different version of what appears to be the same promise. The recipient always assesses whether two promises are equivalent and whether the outcomes are compatible or not. An authority is thus a calibrating agent, in the language of section 4.5.8.

Example 105 (Judges and supreme courts). *Judges at authorities on the law, yet a panel of judges may still disagree about its interpretation. The law attempts to offer (+) clarity on certain situations, but that expression still needs to be accepted and received (-) by agents who view it from different contexts. Each judge is an authority, and the panel of judges can form its own authority as a 'supreme' superagent.*

The interpretations of authority, as power or right, may be sketched as follows:

- *Imposition*: an authority imposes influence on a subordinate:

$$\text{Authority} \xrightarrow{+\text{influence}} \blacksquare \text{Subordinate.} \qquad (13.113)$$

The imposee of the imposition still formally needs to accept the uninvited imposition. If it fails to do so, one can still imagine that the imposer could assert its authority by force. For instance, if the imposer can conquer and subsume the subordinate, so that the agent becomes a part of it, then it can simply promise whatever it likes on its behalf, without violating any notions of autonomy. By absorbing the agent within a larger superagent boundary, its autonomy is lost as far as exterior agents are concerned.

- *Promise:* an authority can be accepted purely as a matter of voluntary cooperation, if granted a mandate M by a number of subjects to issue commands and make decisions C:

$$\text{Subordinate} \xrightarrow{+M} \text{Authority} \qquad (13.114)$$

$$\text{Authority} \xrightarrow{-M} \text{Subordinate} \qquad (13.115)$$

$$\text{Authority} \xrightarrow{+C \mid M} \text{Subordinate} \qquad (13.116)$$

$$\text{Subordinate} \xrightarrow{-C} \text{Authority.} \qquad (13.117)$$

In this way, authority is a symbiotic relationship. Followers promise their support, the authority accepts that and uses it as a basis for making singular decisions conditionally on the mandate, which are then accepted by the followers. This is the basis structure used in democracy. The mandate M may be interpreted as the 'right' or permission to lead (see section 13.7).

In general, a coherent balanced symbiosis might have to be seeded by the imposition of force in order to stabilize a behaviour into a habit initially, which has interesting implications for statecraft and popular moral positions on leadership.

The concept of authority is also widely used, in its derivative meaning, to refer to management roles in systems, where it means the authoritative source of policy decisions—a manager or 'boss'. Authority over other agents is a derivative concept: the promise for which a manager or boss is authoritative is in calibrating policy for other agents to accept and follow—including the promise of wages, which it can withdraw as leverage. Following this agent's policy is, in turn, often assumed to be an obligation: however, based on the foregoing sections, the 'right' to impose commands as a single source of intent must be granted by the sources of its acceptance.

Definition 72 (Authority over other agents). *An agent A to which a number of agents $\{C\}$ have promised to subordinate themselves, by accepting a promise of policy P, given a mandate M:*

$$A \xrightarrow{+P|M} \{C\} \quad (13.118)$$
$$\{C\} \xrightarrow{-P} A. \quad (13.119)$$

The common promise of $-P$ forms the appointment to the position of manager or boss. This appointment to take on the capability of deciding P is what makes the manager an 'authority'. The 'right to manage' is the promise to accept P.

Promise Theory tells us that such authority over other agents is not an inherent 'right' to impose upon them; the authority concept refers only to the integrity of its trusted information. However, the voluntary acceptance of imposition could be interpreted as a mandate (it plays the same formal role). The 'right' to issue commands and directives must ultimately granted by the agent subjects, by accepting the policy—or be overwhelmed by conquering taking over the subject. This has the effect of subordinating the manager to its subjects in return for the privilege of being able to decide policy.

The implications of this are profound: unless a boss or manager's commands can be upheld by overwhelming brute force, or by threat (e.g. suspending wages), then it has to seek this cooperative mandate to play its role as manager[65]. If the manager does not control the policy about wages, then it does not have that leverage, and needs to maintain

13.9. TRUSTED THIRD PARTIES AND WEBS OF TRUST

support by mutual cooperation. A manager or authority is therefore a *role by appointment* (see section 4.4). The manager and subjects are actually coupled in a symbiotic state of mutual subordination, rather than a unidirectional hierarchy of subordination.

Partial authority may also be delegated by a manager to middle managers M, by conditionally promising to follow their decisions on a subset $p_M \in P$.

$$A \xrightarrow{-(p_M \subseteq P)} M. \qquad (13.120)$$

This is a basic application of the matroid pattern (section 4.5.8). The managers now have a mandate to decide matters $p \subseteq p_M$

$$M \xrightarrow{+p|p_M} C, \qquad (13.121)$$

$$M \xrightarrow{+p|p_M} \blacksquare \; A. \qquad (13.122)$$

which now becomes an effective imposition on A It's interesting that, by delegating authority a single source of authority, at the top of a hierarchy, effectively places itself back into a subordinate role, having promised to accept the delegated (though limited) decisions of M, made by the delegates. It thus takes on risk—promising to honour whatever has been delegated. This assumes a further level of trust between the agents: namely that the delegate will not promise more than the authoritative would have done. Thus, with delegated authority by appointment, an organization is in a kind of equilibrium of intent, upheld by trust.

13.9 TRUSTED THIRD PARTIES AND WEBS OF TRUST

Trust is closely associated with promises and information. There are essentially only two distinct models for information distribution: centralization and *ad hoc* epidemic flooding. Alternatively one might call them, central-server versus peer-to-peer.

Example 106. *Two so-called trust models are used in contemporary technologies today, reflecting these approaches: the Trusted Third Party model (e.g. X.509 certificates, TLS, or Kerberos) and the Web of Trust (as made famous by the Pretty Good Privacy (PGP) system due to Phil Zimmerman and its subsequent clones). Let us consider how these models are represented in terms of our promise model.*

The centralized solution to "trust management" is the certificate authority model, introduced as part of the X.509 standard used in web authentication and modified for a variety of other systems (See fig. 13.16)[IT93, Rec97, HPFS02]. In this model, a central authority has the final word on identity confirmation and often acts as a broker between parties, verifying identities for both sides.

A central authority promises (often implicitly) to all agents the legitimacy of each agent's identity (hopefully implying that it verifies this somehow). Moreover, for each consultation the authority promises that it will truthfully verify an identity credential (public key) that is presented to it. The clients and users of this service promise that they will use this confirmation. Thus, in the basic interaction, the promises being made here are:

$$\text{Authority} \xrightarrow{\text{Legitimate}} \text{User} \qquad (13.123)$$
$$\text{Authority} \xrightarrow{\text{Verification}} \text{User} \qquad (13.124)$$
$$\text{User} \xrightarrow{U(\text{Verification})} \text{Authority} \qquad (13.125)$$

To make sense of trust, we look for expectations of the promises being kept.

1. The users expect that the authority is legitimate, hence they trust its promise of legitimacy.

2. The users expect that the authority verifies identity correctly, hence they trust its promise of verification and therefore use it.

Users do not necessarily have to be registered themselves with the authority in order to use its services, so it is not strictly necessary for the authority to trust the user. However, in registering as a client a user also promises its correct identity, and the authority promises to use this.

$$\text{User} \xrightarrow{\text{Identity}} \text{Authority} \qquad (13.126)$$
$$\text{Authority} \xrightarrow{U(\text{Identity})} \text{User} \qquad (13.127)$$

One can always discuss the evidence by which users would trust the authority (or third party). Since information is simply brokered by the authority, the only right it has to legitimacy is by virtue of a reputation. Thus expectation 1. above is based, in general, on the rumours that an agent has heard.

Most of the trust is from users to the authority, thus there is a clear subordination of agents in this model. This is the nature or centralization.

Scepticism in centralized solutions (distrust perhaps) led to the invention of the epidemic trust model, known as the Web of Trust (see fig. 13.17)[AR97]. In this model, each individual agent is responsible for its own decisions about trust. Agents confirm their belief in credentials by signing one another's credentials. Hence if I trust A and A has signed B's key then I am more likely to trust B. As a management approximation, users are asked to make a judgement about a key from one of four categories: i) definitely trustworthy, ii) somewhat trustworthy, iii) un-trustworthy, iv) don't know. An agent

13.9. TRUSTED THIRD PARTIES AND WEBS OF TRUST

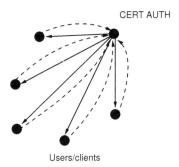

Figure 13.16: The Trusted Third Party, e.g. TLS or Kerberos. A special agent is appointed in the network as the custodian of identity. All other agents are expected to trust this. The special agent promises to verify the authenticity of an object that is shared by the agents. In return for this service, the agents pay the special agent.

then compares these received valuations to a threshold value to decide whether or not a credential is trustworthy to it. The promises are between the owner of the credential and a random agent:

$$\text{Owner} \xrightarrow{\text{Identity}} \text{Agent} \quad (13.128)$$
$$\text{Agent} \xrightarrow{U(\text{Identity})} \text{Owner} \quad (13.129)$$
$$\text{Agent} \xrightarrow{\text{Signature}} \text{Owner} \quad (13.130)$$
$$\text{Owner} \xrightarrow{U(\text{Signature})} \text{Agent} \quad (13.131)$$

The owner must first promise its identity to an agent it meets. The agent must promise to believe and use this identity credential. The agent then promises to support the credential by signing it, which implies a promise (petition) to all subsequent agents. Finally, the owner can promise to use the signature or reject it. Trust enters here in the following ways:

1. The agent expects that the identity of the owner is correct and trusts it. This leads to a Use promise.

2. The Owner expects that the promise of support is legitimate and trusts it. This leads to a Use promise.

What is interesting about this model is that it is much more symmetrical than the centralized scheme.

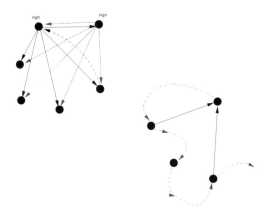

Figure 13.17: In a web of trust an agent signals a promise to all other agents that it has trusted the authenticity of the originator's identity. As a key is passed around (second figure) agents can agree by promising its authenticity, e.g. by signing it or not.

13.10 Leadership in a human system

The subject of leadership is an interesting one. It overlaps with several of the foregoing topics: management, the mandate of authority responsibility, and assurance of outcome. A leader might be characterized as an agent who has assumed responsibility for an outcome, or who has received a mandate to act as an authority is making policy, or one who grants permission for subordinates to act. The basic disagreements about what leadership is supposed to be often revolve around these different interpretations—and, in practice, we speculate that problems of leadership arise because appropriate promises have not been made in an appropriate way.

When planning an outcome, as a leader or manager of a predominantly human system, it's tempting to formulate what we imagine to be a straight path to the outcome—from where we are to the outcome we desire. That path might not be the path of least resistance. Indeed, it might not even be possible to construct[66]. A project or outcome planner needs to explore what paths *can be promised*. Many management frameworks lay down processes to try to apply brute-force persuasion and resources to solve problems. From a Promise Theory perspective, this doesn't make sense. The alternative is to start by surveying what promises are currently given, then what promises might be given if invited in the right way. The human scenario is very different from a machine scenario.

Leadership may be a promise to lead, based on promises of a mandate of authority—but what tools should an authoritative agent use to maintain the fragile symbiotic cooperation of a group? An approach that works on a small scale might not work on a large scale, as we know of phase transitions, yet leaders nearly always try to maintain the same

13.10. LEADERSHIP IN A HUMAN SYSTEM 191

approach as companies and nations grow. Not thinking through the actual promises made may leave us with no idea about why an outcome failed. Many an outcome failed because of improper assumptions—not doing one's homework to find the actual state of affairs. Promises express semantics but also context, in the form of a network of relationships and dependencies. There is key information in those networks that tells the channels of cause and effect—how influence propagates through a system, like a human community.

Example 107 (Undifferentiated leadership as a job rather than a role). *Leaders often believe that those agents structurally under them in an organizational hierarchy have granted a mandate for them to subordinate themselves voluntarily in all matters, when in fact no such mandate has been given. Organizational structure is a blunt instrument, which doesn't normally distinguish authority based on specialization—rank in a hierarchy supercedes genuine authority. As a result, many agents may feel attacked by impositions on matters where they believe the leader agent exceeds its authority. Job title is not always a sufficient discriminator to mandate authority on a wide range of issues.*

CHAPTER 14

COMPONENTIZATION AND MODULARITY IN SYSTEMS

The design of functional systems from components is a natural topic to analyze using promises: how we break down a problem into components, and how we put together new and unplanned systems from a pre-existing set of parts.

Decomposing systems into components is not *necessary* for systems to work (an alternative with similar cost-benefit is the use of repeatable patterns), but it is often desirable for economic reasons. A component design is really a commoditization of a reusable pattern, and the economics of component design are the same as the economics of mass production.

Example 108. *The electronics industry is a straightforward and well known example of component design and use. There we have seen how the first components: resistors, inductors, capacitors, and transistors were manufactured as ranges of separate atomic entities, each making a variety of promises about their properties. Later, as miniaturization took hold, many of these were packaged into 'chips' that were more convenient, but less atomic components. This has further led to an evolution of the way in which devices are manufactured and repaired. It is more common now to replace an entire integrated unit than to solder in a new capacitor, for example.*

Example 109. *An analogy with medicine would be that one favours transplanting whole organs rather than repairing existing ones.*

Example 110. *For example, in a particular context, an electrical appliance might be modelled as an agent that makes some kind of promise, perhaps to boil water or play* music. However, we also know that it is built from smaller electrical components that can be purchased separately, each making their own promises, and used to make any number of different circuits (coils, resistors, capacitors, etc). These could also be modelled as agents if we care for that level of detail. Electronic design is a showpiece for other industries in regard to component based design. Furniture manufacturers, currently exemplified by IKEA, have imitated this approach to sell furniture built from components like wood, screws, metal brackets etc.

14.1 Definition of components

A component is an entity that behaves like a promise-agent and makes a number of promises to other agents, in the context of a system. It can be represented as a bundle of promises. Since a component must fit together with other components, it is likely to be a parameterized object in order to adapt to different patterns of use.

Let us define components within the scope of a larger system. For weakly coupled systems, components can be called standalone.

> **Definition 73** (Standalone or independent components). *A standalone or independent component is an agent that makes only unconditional promises.*

In strongly coupled systems or subsystems, components are integral parts, embedded in a larger whole.

> **Definition 74** (Embedded or dependent components). *A dependent component is an agent that makes one or more conditional promises.*

14.2 What systemic promises should components keep?

The kind of promise a component makes can be used to say something about its properties. In addition to the functional promises that a component would try to keep, there are certain issues of design that would tend to make us call an agent a component of a larger system. Typical promises that components would be designed to keep could include:

- To be replaceable.
- To be reusable.
- To behave predictably.

- To be (maximally) self-contained, or avoid depending on other components to keep their promises.

What about the internal structure of agents in the promises? Clearly components can have as much private, internal structure as we like – that is a matter of design choice.

14.3 Component design and roles

The simple structure of promise patterns allows us to map components simply onto roles. A component is simply a collection of one or more agents that forms a role, either by association or a cooperative role.

Example 111. *Components and their roles in some common systems*

System	Component	Role
Television	Integrated circuit	Amplifier
Pharmacy	Pill	Sedative
Patient	Intravenous drug	Antibiotic
Vertebrate	Bone	Skeleton
Cart	Wheel	Mobility enabler
Playlist	Song by artist	Soft Jazz interlude

14.4 Promise conflicts in component design

Building an appliance on components that make exclusive promises, i.e. promises that can only be kept to a limited number of users is a fragile strategy. Components that make exclusive promise experience the issue of conflicts of promise usage (see section 8.2). It can be considered good practice to try to design for the avoidance of such conflicts, however one often has incomplete information about the usage of components, making this impossible. This is one of the flaws in top-down design.

> **Definition 75** (Contentious components). *Components that make conditional promises, where the condition must be satisfied by only one exclusive promise from another component may be called* contentious. *Multiple agents contend with one another to satisfy the the condition, but only one can do so.*

Example 112. *It is only possible to plug one appliance into a power socket at a time.*

Example 113. *Only one patient can swallow a pill at a time, thus a pill is a contentious agent that promises relief if taken.*

14.4. PROMISE CONFLICTS IN COMPONENT DESIGN

Example 114. *Integrated circuitry, for instance, has been a major success in electronics. However, independence from internal structure can also be a strength. The more elementary the parts, the more ways there are to combine them into a creative chemistry. If we design components to rely on other components external to themselves, we haven't done anything fundamentally bad. But, if we design a component to have a mandatory dependence on a shared, exclusive resource then we have introduced contention for the resource. That increases the likelihood that the decision will lead to design problems in the long run.*

Designing with contentious components can sometimes save on costs, but it can lead to bottlenecks and resource issues in continuous system operation. Contentious promises are therefore fragile.

Example 115. *An engine that promises to run if there is a special kind of fuel.*

Example 116. *Multiple electrical appliances connected to the same power line contend for the promise of electrical power. If the maximum rating is a promise of, say 10 Amperes, they multiple appliances requiring a high current could easily exceed this.*

Example 117. *Software applications running on a shared database or server have to contend for availability and capacity.*

The foregoing examples show the effect of imperfect information during component design. The designer of an appliance cannot know the power or server availability when designing appliances. A similar example can be made with traffic.

Example 118. *Cars could be called contentious components of traffic, as they contend for the limited road and petrol/gas pumps at fuelling stations.*

A deliberate choice is often made to minimize costs, given the probability of needing to deliver on a promise of a resource.

Example 119. *Computer racks are computers found in datacentres. Normally they do not have their own screens or keyboards, as these are rarely needed. A single keyboard and screen and DVD player might be shared between many of them. These blade computers may be called contentious as they potentially expect to be able to use the common keyboard and screen at any time, even though this is not possible.*

It is worth remarking that contention is also closely related to coordination and calibration (see chapter 5), as a single point of coordination is also a single point of contention and indeed failure. Contention leads to fragility.

14.5 REUSABILITY OF COMPONENTS

Let us suppose our goal is to engineer components so that the promises they keep allow the components to be reused in different scenarios. There might be cases where it doesn't matter which of a number of alternatives is chosen – the promises made are unspecific and the minor differences in how promises are kept do not affect a larger outcome. In other scenarios it could matter explicitly.

Example 120. *Imagine two brands of chocolate cake at the supermarket. These offerings can be considered components in a larger network of promises. How do we select or design these components so that other promises can be kept?*

- *Doesn't matter: You would normally choose a cheap brand of cake, but it is temporarily unavailable, so you buy the more expensive premium version.*

- *Matters: One of the cakes contains nuts that you are allergic to, so it does not fulfill your needs.*

If the promise made by the cake is too unspecific for its consumer, there can be a mismatch between expectation and service.

14.5.1 DEFINITION OF REUSABILITY

Reusability is achieved when multiple consumers' expectations, documented by what use-promises they are willing to make, can be quenched by a component's offer in a *conditional promise binding*.

Definition 76 (Reusability). *Let $C_i, i = 1, 2, \ldots$ be a set of components that make conditional promises π_i, with bodies $b_i | d$, that depend on using a component supplying d, delivered by a mutual component C_D. We say that C_D is a reusable component amongst $\{C_i\}$ if and only if*

$$\begin{aligned} \{C_i\} &\xrightarrow{U_i(d)} C_D \\ \{C_i\} &\xleftarrow{d} C_D \end{aligned} \tag{14.1}$$

where $U_i(d) \subset d$, for every i. In other words, a component will fit into each one of these scenarios if it meets or exceeds all the requirements to use its promises.

A component cannot be intrinsically reusable, but only reusable relative to the other components in its system of use.

Reusability does not necessarily affect the perceived value of a component (see section 9.1). An agent might make do with a component offering most of its desired

14.5. REUSABILITY OF COMPONENTS

promises, if there are other non-functional reasons for selection within the scope of a model.

Example 121. *Suppose a computer has printer driver version 12 as a software component, and three software components use this version of the printer driver. The spreadsheet component only promises to print if version 13 or greater of the printer driver is available. The drawing software component only works with version 4 of the driver, or if a separate printing program is installed. However, a permissive calendar program has few requirements and will work with any, because it only uses a small number of the features of the driver.*

14.5.2 INTERCHANGEABILITY OF COMPONENTS

When components are changed, or revised, the local properties of the resulting system must change. This might further affect the promises of the entire system, as perceived by a user, if the promise bindings allow it. In this case, either the total system has to change its promises, or a total revision of the design might be needed to maintain the same promises of the total system.

> **Definition 77** (Interchangeability/equivalence of components). *Two components C_1 and C_2 are interchangeable, i.e. equivalent with one another, relative to an agent A, if and only if every promise made by C_1 and used by A is identically made by C_2, or vice versa.*

Components may be assessed to be compatible or interchangeable by any agent that is a potential user.

Example 122. *A radio jockey consuming music for a show assesses musical pieces according to the style and artist's promises to decide whether they are compatible with the show's promises to listeners. Listeners to the show form their own assessment in turn.*

Example 123. *A knife can be interchanged with an identical new knife in a restaurant without changing the function, even if one has a wooden handle and the other plastic.*

Example 124. *A family pet cannot be interchanged with another family pet without changing the total family system.*

14.5.3 COMPATIBILITY OF COMPONENTS

We define compatibility of components as a weaker condition than interchangeability. If components are compatible, it is possible for them to connect with one another, even if the precise promises are not maintained at the same levels.

198   CHAPTER 14.   COMPONENTIZATION AND MODULARITY IN SYSTEMS

> **Definition 78** (Compatibility of components). *A component is compatible with another if its ± promises can form all the same bindings with neighbouring components, without necessarily preserving all promised details.*

Example 125. *Viruses interface with cells or other systems because the systems have promises that are compatible with the virus, although the function of the total system is perverted. Antibodies recognize antigens like viruses because they have receptors that are compatible with the molecular signatures of the viruses[67]. We recognize smells because certain molecules are compatible with receptors in our noses.*

Example 126. *British power plugs are not interchangeable or compatible with the power sockets in America or the rest of Europe because they do not make complementary promises. Similarly, we can say that British and American power plugs are not interchangeable.*

14.5.4   BACKWARDS COMPATIBILITY AND REGRESSION TESTING

Replacement components are often said to require backwards compatibility, meaning that a replacement component will not affect the promises made by an existing system. In fact, if one wants status quo, the correct term should be interchangeability, as a new component might be compatible but still not function in the same way.

When repairing devices, we are tempted to replace components with new ones that are almost the same, but this can lead to problems, even if the new components are 'better' in relation to its own promises.

Example 127. *The promise of increased power in a new engine for a plane might lead to an imbalance in the engines, making handling difficult. Similarly, changing one tyre of a car might cause the car to pull in one direction.*

Making all the same promises as an older component is a necessary condition for interchangeability, but it is not sufficient to guarantee no change in the total design. We need identical promises – no more and no less, else there is the possibility that new promises introduced into a component will be able to conflict with the total system in such a way as to be misaligned with the goals of the total design.

Example 128. *Diesel fuel makes many of the same promises as petrol or gasolene, but these fuels are not interchangeable in current engines.*

Example 129. *When fingerprint readers were installed on to PCs in the beginning, they allowed users to choose between typing a password or using a fingerprint to enter the system. The upgraded keyboard added a promise which weakened security as allowing*

14.5. REUSABILITY OF COMPONENTS

one or the other, rather than requiring both at the same time gave two possible ways of attacking the system instead of only one.

Example 130. *When pluggable components were added to PCs during the 2000s, allowing upgrading to faster CPUs, the upgraded chips used more power and generated more heat. The power and cooling requirements thus changed and sometimes PCs would overheat.*

Example 131. *The insertion of new output transistors with better specification into a particular HiFi amplifier actually improved the sound quality of the device. The transistors cannot be considered interchangeable, even though there was an improvement, because the promises they made were not the same, and the total system was affected. Now the older system was weaker and cannot be considered an adequate replacement for the new.*

When upgrading components that are not identical, a possibility is to design in a compatibility mode such that no new promises are added when used in this way. This allows component promises to remain interchangeable and expectations to be maintained. Downgrading can also be compatible but not interchangeable.

Definition 79 (Regression test). *An assessment of a component's promise, as probed by a use-promise that measures at a constant level of expectation.*

$$A_{\text{tester}} \xrightarrow{-b_t} \{C_i\}. \tag{14.2}$$

where i runs over all component versions. If the promise can be kept, the test is passed.

Theorem 3 (Regression test limitation). *The ability to detect faults in components is limited to those precise behaviours b the test-agent promises to use with by promising $-b$.*

The proof follows from the discussion in section (3.11.7). Let the promise body of component C_i be b_i. The effective action of testing is to measure the overlap $b_i \cap b_t$ between the components and the test b_t. New features cannot be in b_t, since they are not expected. Hence the test cannot promise to find them.

14.5.5 GENERIC INTERFACE EXPECTATIONS

How does an agent select one specific component amongst many similar choices?

Suppose we have a number of components that make similar promises. If the promises are identical, then the components are identical and it clearly doesn't matter which component we use. However, if the promises are generic and vague, there is a high level of uncertainty in forming expectations about the behaviour of the components (see section 14.6.2).

Example 132. *Suppose a car rental company promises us a vehicle. It could give us a small model, a large model, a new model, a clapped-out wreck or even a motorbike. It could be German, Japanese, American, etc. Clearly the promise of 'some kind of vehicle' is not sufficiently specific for a user to form useful expectations.*

Example 133. *In a restaurant, it says 'meat and vegetables' on the menu. Does this mean a medium-rare beef steak with buttered carrots and green beans, or hamburger and fries?*

The potential for a mismatch of expectation steers how components should design interfaces to systems with similar promises. It is often desirable for an agent consuming promises to have a simple point of contact (sometimes called a 'front end') that gives them a point of 'one stop shopping'. This has both advantages and disadvantages to the user, as interchangeability of the back-end promises is not assured.

Example 134. *In a food court, the different food stalls offer meals for five dollars a piece with hundreds of choices. You pay at the same checkout, but the stalls offer very different food.*

In a gourmet restaurant, the experience of choice is simplified to a 'taster menu', which the chef decides each night. There are some basic choices for special diets and allergies, but the simplification is based on the trust of the components designed by the chef.

If a promiser makes a 'lowest common denominator' promise, it is easy to consume, but one does not trust the promise.

Example 135. *Supermarkets often re-brand their goods with generic labels like 'mega-store beans', or 'super-saver ice cream'. The lack of choice makes it easy for the user to choose, but it also means the user doesn't know what they are getting. Often in these cases, the promise given is implicitly one of a minimum level of quality.*

Example 136. *In computing, a service provider offers a 'SQL database' service. There are many components that offer this service, with very different implementations. PostgreSQL, MySQL, MSSQL, Oracle, and several others all claim to promise an SQL database. However, the detailed promises are very different from one another so that the components are not interchangeable. Writing software using a common interface to these databases thus leads to using only the most basic common features.*

14.5. REUSABILITY OF COMPONENTS

> **Law 12** (Lowest common denominator). *When a promise giving component makes a generic promise interface to multiple components, it is forced to promise the* least common denominator *or minimal common functionality. Thus generic interfaces lead to low certainty of outcome.*

The user of a promise is the driver and the filter to decide what it needs to satisfy, i.e. the agent's *requirements*[68]. If the user promises to use any 'meat and vegetables' food promise by a caterer, he will doubtless end up with something of rather low quality, but if she only promises to use 'seared filet, salted, no pepper', etc, the provider will have to promise to deliver on that, else she would not accept it to form a promise binding.

When designing or productizing components to consumers, one often uses the language of imposing requirements.

> **Law 13** (User expectations). *Use-promises or impositions of specification play equivalent roles in setting user expectations for promised outcomes. See sections 3.11.5 and 5.5.1.*

One sees that a use-promise that seeking a matching binding, i.e. to shop between alternatives, needs to be sufficiently specific in its expectations or requirements to prevent bindings from being made too easily and failing to meet expectations. In other words, agents need to be sufficiently discerning.

14.5.6 IRREVOCABLE COMPONENT CHOICES

Selecting one particular component from a collection has long term consequences for component use and design if one selects a non-interchangeable component. The differences that are specific to its promises then lead the system along a path that moves farther away from the alternatives because other promises will come to rely on this choice through dependencies.

Example 137. *Choosing to use a fuel type to power a car, e.g. using a car that makes diesel rather than a petrol/gasolene promise has irreversible consequences for further use of that component. Choosing the manufacturing material for an aircraft, or the building site for a hotel has irreversible consequences for their future use.*

Example 138. *Choosing to use one political party has few irreversible consequences, as it is unlikely that any other promises by the agent making the choice would depend on this choice.*

14.5.7 VERSIONING, EVOLUTION AND PROMISE CONFLICTS

Designers matching the evolution of a system, through progressive versions of its components, to the evolution of its users' expectations. This evolution happen can happen slowly or quickly. If there is a mismatch between the rate of change of expectations and versions, then providers and consumers rarely come into equilibrium.

Consider figure 14.1, which shows how a single promise agent can depend on a number of other agents. If the horizontal stacking represents components in a design, and vertical stacking indicates different versions of the same component, then each agent would be a replaceable and upgradable part.

In a system with multiple cross-dependencies, different versions of one component may depend on different versions of others. How may we be certain to keep consistent promises throughout a total system made of many such dependencies, given that any of the components can be re-used multiple times, and may be interchanged for another during the course of a repair or upgrade. There is a growth of *combinatorial complexity*.

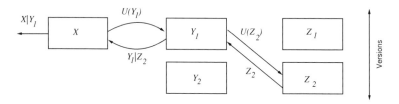

Figure 14.1: A chain of component dependencies horizontally, and changing versions vertically, leads to combinatoric promise inflation. The cost of promising all these dependences can be prohibitive for system design.

Consider the component arrangement in figure 14.1, showing a component X that relies on another component Y, which in turn relies on component Z. The components exist in multiple versions, and the conditional promises made by X and Y specifically refer to these versions, as they must in order to be certain of what detailed promise is being kept (as in section 14.5.5).

This situation is fragile if components Y and Z are altered in some way, there is no guarantee that X can continue to promise the same as before, thus its promise to its users is conditional on a given version of the components it relies on, and each change to a component should be named differently to distinguish these versions (see section 14.5.8 below).

To avoid this problem, we can make a number of rules that guide the design of components.

14.5. REUSABILITY OF COMPONENTS

Rule 9 (Component dependency can be made version specific). *Promises by one component to use another can avoid incompatible promises by referring to the specific versions that quench their promises to use. This avoids the risk of later invalidation of the promises by the component as the dependency versions evolve.*

Versioning takes care of the matter of component evolution and improvement in the presence of dependency. The next issue is what happens if a component is used multiple times within the same design. Can this lead to conflicts? There are now two issues surrounded repeated promises:

- If multiple components, with possibly several versions, make completely identical promises, this presents no problem, as identical promises are idempotent (see section 3.13).

- Incompatible promises, or exclusive promises prevent the re-use of components.

Rule 10 (Reusable components and avoidance of exclusive promises). *Any component that can be used multiple times in the same system must be able to exist along side other versions and patterns of usage. This means that components should not make conflicting promises across versions, if multiple versions are to coexist.*

Example 139. *Suppose a castle has one flag for signalling. The guards promise to raise the flag if the king is at home, and lower it if he is absent. The courtier promises to raise the flag only on the king's birthday. The flag is re-usable, but not at the same time.*

Example 140. *The English word bow promises multiple meanings (it is a homonym). It means:*

- *An implement used to play instruments such as the violin*
- *To bend forward at the waist in respect (e.g. "bow down")*
- *The front of the ship (e.g. "bow and stern")*
- *The weapon which shoots arrows (e.g. "bow and arrow")*
- *A kind of knotted ribbon (e.g. bow on a present, a bow-tie)*
- *To bend outward at the sides (e.g. a "bow-legged" cowboy)*

Most of these usages have distinct contexts and pronunciations. If a cowboy was ordered to bow, would he bend his legs outwards or double over?

The unique naming of components is a simple way to avoid some of these problems. We shall describe this in the next section.

14.5.8 Naming and 'branding' of components

The naming of components turns out to be a highly important issue for identifying function in a changing environment. Components need to to signal their promised function somehow, within their scope. Naming allows components to be recognizable and distinguishable for reuse. Naming need not be done through language; recognition may occur through shape, texture, or any form of sensual signalling. In marketing, this kind of naming is often called *branding*[69].

It makes sense that bundles of promises make naming agreements within a system, according to their functional role, allowing components to be

- Quickly recognizable for use.
- Distinguishable from other components.

Because an agent cannot impose its name on another agent, nor can it know how another agent refers to it, naming conventions have to be *common knowledge* within a collaborative subsystem. Establishing and trusting names is one of the fundamental challenges in systems of agents[70].

Naming needs to reflect variation in time and space, like virtual locational coordinates. Variation in space can be handled through scope, as it is parallel. However, variation in time is a serial property, indistinguishable from versioning, and therefore naming of a set of promises to be unique within a given scope.

> **Definition 80** (Naming or identity of component function (+ promises)). *A component name or identifier is any representative promisable attribute, rendered in any medium, that maps uniquely to a set of two promises:*
>
> - *A description of the promises kept by the component (e.g. text like 'Ford Escort', or a shape like handle indicating carrying or opening)*
> - *A description of model or version in a family or series of evolving designs (e.g. 1984 model, version 2.3, "panther")*

We often keep a version identifier separate from the functional name because these things can be promises separately, and thus promise principles indicate that these are two separate promisers for the component.

Example 141. *In information technology, a user's public key is a name that represents it uniquely once trusted by other users.*

Example 142. *In computer operating systems, software is usually componentized into 'packages'. These packages are named with a root identifier and a version number, e.g.*

14.5. REUSABILITY OF COMPONENTS

```
flash-player-kde4-11.2.202.243-30.1.x86_64
unzip-6.00-14.1.2.x86_64
myspell-american-20100316-24.1.2.noarch
kwalletmanager-4.7.2-2.1.2.x86_64
```

The root name leads up to a number. Notice that the version numbering needs only to be unique for a given component, as each component essentially lives in its own 'world' or subsystem, designated by the root name. The number then refers to temporal version, and a suffix x86_64 *or* noarch *refers to version in 'space', or on different computer architectures.*

Within any particular scope, the user of components would normally see value in the uniqueness of names. We shall discuss this further in section 14.6. For now, we note that names need not be unique.

Definition 81 (Component alias). *Any name that represents a component with identical promises to another component may be considered an alias for the name of the component.*

14.5.9 NAMING COMPONENT USAGE OR PROMISE CONFIGURATIONS

We may also decide to name individual applications of components, i.e. name a particular configuration in which a component is used. These usage cases need not be unique, as promises are idempotent so repeated use of a component would not make any difference to the formalism. In the physical world, however, a realization of a component can normally only be used once.

Definition 82 (Naming or identity of component usage (− promises)). *A description of what input data, parameters or arguments are being supplied to the component in a use-case involving the promise components.*

Attaching distinct names to instances of usage can be useful for identifying when and where a particular component was employed, e.g. when fault finding.

Example 143. *In computer programming, generic components are built into libraries that expose promises to using patterns of information. This is sometimes called an API or Application Programmer Interface. The application programmer, or promise user, makes use of the library components in a particular context for a particular purpose. Each usage of a particular component may thus be identified by the information passed to it, or the other components connected to it. We may give each such example a name* *'authentication dialogue', 'comment dialogue', etc.* We can also use the set of parameters given to an API interface promise as part of the unique name of the reference.

If names represent functionality, then they must also reflect changes to that function. This is the purpose of versioning. Even if the goal of a particular component is constant (e.g. detergent should clean), the evolution of component design will normally lead to a succession of versions in an attempt to improve the components by various value-judgement criteria (see chapter 9).

Naming different patterns of usage according to definition 80 has the advantage of allowing different components to address one another specifically, without muddling their worlds. We summarize with another rule of thumb.

Rule 11 (Uniqueness of naming ± promise agents). *The unique naming of agents, operating in the role of non-redundant components, is both necessary and sufficient to avoid ± promise conflicts, within a certain scope. With unique naming, all components may co-exist.*

14.6 THE ECONOMICS OF COMPONENTS

When does it make economic sense to build components? Each promise carries an expected value, assessed individually by the agents and other components in a system.

14.6.1 THE COSTS OF MODULARITY

Even from the viewpoint of a single agent, the cost associated with choosing a modular design is not necessarily a simple calculation to make. It includes the on-going collaborative cost of isolating components into weakly coupled parts, designing their interfaces, and also the cost of maintaining and evolving the parts over time. The accessibility of the parts adds a cost too. If components are hard to change or maintain, because they are inside a closed box, then it might be simpler to replace the whole box. However, if the component is cheap and the rest of the box is expensive in total, then the reckoning is different.

Example 144. *Changing a flat tyre rather than replacing an entire car makes obvious sense, because the tyres are accessible, large and expensive enough to be worth the effort.*

Example 145. *Electronic components like transistors and resistors can be replaced individually, so it makes sense to change failed components individually, thus saving the cost of unnecessary additional replacements. However, the miniaturization of electronics eventually changed this situation. At some point it became cheap to manufacture integrated*

14.6. THE ECONOMICS OF COMPONENTS

circuits that were so small that changing individual subcomponents became impractical. Thus the economics of componentization moved towards large-scale integration of parameterizable devices.

Example 146. *Virtual computational components, like software running on virtual servers in a datacentre, are apparently cheap to replace, and are used extensively to manage the evolution of software services. However, while these components are cheap to build, existing components gather history and contain runtime data that might be expensive to lose.*

Modularity is general considered to be a positive quality in a system., but when does componentization actually hurt? There are many ways we might assess this. We can begin by following basic Promise Theory structural principles:

- Flaw in (+) promise: i.e. if the components are not properly adapted to the task they support, because they don't make the right promises, perhaps due to poor understanding of the larger context and patterns of usage.

- Flaw in (-) promise: i.e. if the integration of the components into a larger design is not optimal so that the value of the components is never utilized fully. This includes under- and over-qualification for the task.

- Cost of (+): when the cost of making, packaging, and maintaining the components exceeds the cost of just building the total system, over the expected lifetime of the system.

- Cost of (-): if the additional overhead of connecting with the components (combinatorics) exceeds other savings.

- Conditional promises: early modularization might constrain one's ability to innovate, because it changes the nature of the network of promises from being free (as hoc) to being constrained across certain boundary interfaces. Early choices about the balance between interior details and exterior promises (interfaces) can set the stage for costly redesign later.

Example 147 (Microservices). *In software engineering for the cloud computing age, there is a drive to modularize software as multiple service agents that make a loose web of promises to one another. This is contrasted with the idea of a 'monolithic system' in which there is a single agent, tightly integrated. We can use promise theoretic principles to assess whether a microservice (componentized) design is better adapted to its purpose than a monolithic system, or is merely wasteful, on a case by case basis.*

14.6.2 ADDENDUM TO CHOOSING BETWEEN ALTERNATIVE COMPONENTS: FITNESS FOR PURPOSE

The semantics of component selection were discussed in section 14.5.5. As an addendum, we can add that economic valuations of promises can also affect the selection of a particular component.

Example 148. *In sales, one often ends up selling every possible feature in an attempt to differentiate a promise and grab the attention of a buyer.*

The permissiveness of promise expectations referred to in the reusability law is an expression of *fault tolerance*, and this suggests another perspective on reusability of components: quality and 'fitness for purpose'. Can we accept any old promise from a component dependency in a system?

Example 149. *Would we choose a postal delivery that took four weeks instead of four days? Would we accept balsa wood wings on an airliner instead of steel, or smaller screws that could handle smaller loads? Would we accept a less powerful computer in place of another to handle business transactions?*

All of these questions imply that there is a value associated with the promises made by different components, and that some promises are worth more than others. Naturally, this is why premium fuel costs more than standard, and hand-made luxury cakes cost more than mass-produced factory cakes. The evaluation of fitness for purpose is driven by the user and not the giver of a promise.

Marketing, camouflage and imitation are all strategies used to create deceptions around fitness for purpose[71].

CHAPTER 15

PROMISES AND MATHEMATICAL GAMES

15.1 RELATIONSHIP BETWEEN GAMES AND PROMISES

The mechanism of rational, voluntary decision-making used in economics is the theory of games[Ras01, Mye91]. There is a natural relationship between Game Theory and certain constellations of promises. We confine our discussion here to normal or strategic form games[72].

A game is a number of players A_i, where $i = 1 \ldots N$, a set of actions, payoffs and information. A player A_i's *action set* is the entire set of actions or pure strategies available to him[73]. An action combination is a set of one action per player s_i in the game. Player A_i's payoff or utility is denoted $U_i(s_1, s_2, \ldots, s_N)$ and has one of two interpretations:

- It is the utility received at the end of a trial by a player, based on what the player did.

- It is an estimated or expected utility to be considered as a function of the possible or expected strategies a player might choose in a future game.

It is this second interpretation that most closely relates to promises, since we shall identify promises with expectations of actions. However, we should be clear that the latter cannot exist without some measure of the former. This duplicity of interpretation mirrors a similar one in Promise Theory, in which one can measure the reliability of a promise with a scalar probability value. The probability can then be regarded as a history of a player, or as a prediction of future behaviour.

A *strategy* is a rule that tells player A_i which action s_i to choose at each instant in the game, given his or her information set. An equilibrium $s^* = (s_1, s_2, \ldots, s_N)$ is a strategy combination consisting of the 'best' strategies for each player, with respect to the expected payoff. Several such equilibria might exist.

There is a class of games in which the possible actions or moves of a player are the same in each round of a game. Thus each round of an extended game is indistinguishable from multiple games, except perhaps in the information available available to the players. Iterated games are special because both *each round* and the entire *iterated game* can be represented as a utility matrix of the same form. Thus we may discuss both extended interactions and one-shot interactions using a utility matrix formulation as long as we restrict attention to such games. We shall do this here.

Promises are static snapshots of policy. A promise graph such as that in fig. 15.1 can be viewed in two ways: either as an equilibrium over a complete interaction epoch

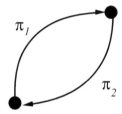

Figure 15.1: A simple exchange of promises.

or as a single round in an repeated interaction. To make the move from games with actions to games with promises, we simply replace actions with promises. Any ordering of promises (who makes which promise first) is analogous to an ordering of actions. In a game one considers a strategy to be a choice of actions. Here we shall assume it to be a set of promises made:

$$A_i \xrightarrow{\{\pi\}_i} A_j \qquad (15.1)$$

where $\{\pi\}$ is potentially a bundle of different promises made from one player to another. This bundle might also comprise the empty set.

The payoff or utility received by the ith player or agent in the system is the sum of

15.2. CLASSIFICATION OF GAMES

its own private valuations of promises made and received:

$$U_i(\{\pi\}_1, \{\pi\}_2, \ldots, \{\pi\}_N) = \sum_{j \neq i} v_i \left(A_j \xrightarrow{\{\pi\}_j} A_i, A_i \xrightarrow{\{\pi\}_i} A_j \right) \quad (15.2)$$

Any kind of promise can be evaluated, regardless of whether it is a service promise or a use-promise[Bur05] (give or take promise).

15.2 CLASSIFICATION OF GAMES

Games are classified in various ways, particularly with regard to cooperation and conflict. An interaction is *cooperative* if players coordinate their actions in order to improve their final payoff (utility). Cooperation might be contingent on available information. An interaction is a *conflict* if the actions or intentions of one player can adversely affect the payoff (utility) of another player's actions. Let us give some examples of these representations of games in terms of promise exchanges.

15.2.1 COOPERATIVE, NO CONFLICT

Consider two agents helping one another to provide a sufficient level of service for a third party. By helping one another, they are able to work together to meet 3's requirements and secure contracts with the third party and hence both profit: they are not in competition with one another. This scenario is shown in fig. 15.2.

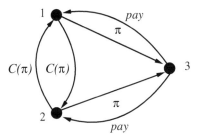

Figure 15.2: Cooperative game, no conflict.

Two agents 1 and 2, make coordinated promises to serve node 3. The third party promises to reward both agents for their service, independently. The values of the remuneration promises $v_{1,2}(pay)$ are not related.

15.2.2 COOPERATIVE, WITH CONFLICT

Consider now two agents who bargain with one another directly for services from one another. Both need these services from the other, so the services are valuable (fig 15.3). By making each other's promises contingent on what they receive from the other, the value each of them receives becomes moderated by the other player. If they cooperate with one another, they can both win a fair amount, but each change can have negative consequences through the feedback. Agent 1 promises π if the other promises ρ and vice

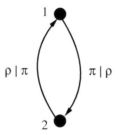

Figure 15.3: Cooperative game, with conflict.

versa. The values $v_1(\rho|\pi)$ and $v_2(\pi|\rho)$ are inter-dependent.

15.2.3 NON-COOPERATIVE, NO CONFLICT

Two agents providing a service in a free market promise the same service to a third party, without any cooperation (fig. 15.4). The third party makes its choice and accepts the promised service partly from one and partly from the other. Each is promised a payment. Again, the value of the payment promises are quite independent, hence there is no conflict between agents 1 and 2.

15.2.4 NON-COOPERATIVE, WITH CONFLICT

This example is the classic and well-known Prisoner's Dilemma (fig. 15.5). Two agents promise to talk (or not talk) to their prison keeper (a third party). The prison keeper rewards both agents depending on the combined promises of both of the agents. The agents 1 and 2 are not in a position to cooperate (being held in different cells), but they affect one another's outcomes indirectly. This is a conflict because the actions of each player can have consequences for both of them, and it is non-cooperative because the two players cannot communicate.

15.2. CLASSIFICATION OF GAMES

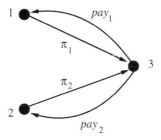

Figure 15.4: Non-cooperative game, no conflict.

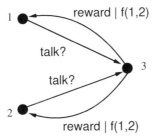

Figure 15.5: Non-cooperative game, with conflict.

15.2.5 CONSTANT-SUM GAMES

Let us now show that we can represent the special case of constant-sum (usually zero-sum) games, provided a certain level of cooperative infrastructure is in place. Such an infrastructure is often assumed in economic texts, but here we must be more careful. A constant-sum game is special because it implies an ability for one agent to control the value received by another agent. In a world of complete autonomy, that is forbidden, since each agent makes its own value judgements. Hence, to understand such a scenario, we must secure certain agreements between agents to agree on a currency standard. Constant sum games are not dependent on centralization, but we present two versions of such a game with and without centralization, both of which however show the fragility of the concept.

Theorem 4 (Constant-sum game). *A promise interaction that forms a constant-sum game requires a finite number (at least N) of agents (N players and at least one adjudicator which may optionally be taken amongst the players), and all agents must agree on a standard currency.*

Proof. Consider N promise interactions of the form:

$$A_i \xrightarrow{\pi_i} A$$
$$A \xrightarrow{r_i|\chi(\pi_1,\pi_2,\ldots,\pi_N)} A_i. \qquad (15.3)$$

Each agent A_i promises an adjudicator A some service or behaviour π_i, and the adjudicator promises, in return, a piece of a common reward r_i, subject to a constraint $\chi(\pi_1, \pi_2, \ldots, \pi_N)$. We would like this constraint to be that the sum of the values perceived by the agents is constant, i.e. we would like

$$\chi = \sum_{i=1}^{N} v_i(A \xrightarrow{r_i|\chi(\pi_1,\pi_2,\ldots,\pi_N)} A_i) = \text{const.} \qquad (15.4)$$

However, each agent's valuation function is its own private estimate and is unknown to A, so the best it can do is to arrange for

$$\chi = \sum_{i=1}^{N} v_A(A \xrightarrow{r_i|\chi(\pi_1,\pi_2,\ldots,\pi_N)} A_i) = \text{const.} \qquad (15.5)$$

Equations. (15.4) and (15.5) can only agree if each agent A_i makes an agreement to use the same valuation $v_i = v_A(A_i) = r_i, \forall i$, implying that the agents have a common currency. In a centralized framework, this implies an agreement of the form

$$A \xrightarrow{v_A} A_i$$
$$A_i \xrightarrow{U(v_A)} A \qquad (15.6)$$

between each node and the adjudicator. Hence the payoff matrix for the game is:

$$U_i \left(v_i (A_i \xrightarrow{\pi_i} A, A \xrightarrow{r_i|\chi} A_i) \right)$$
$$= U_i \left(v_A (A_i \xrightarrow{\pi_i} A, A \xrightarrow{r_i|\chi} A_i) \right)$$
$$= U_A \left(v_A (A_i \xrightarrow{\pi_i} A, A \xrightarrow{r_i|\chi} A_i) \right). \qquad (15.7)$$

The adjudicator node has the role of book-keeper. There is no impediment to identifying A as any *one* of the A_i agents, the conditions are binding and sufficient regardless of which agent plays the role of A. Similarly, agents with the role of A can be repeated any number of times. □

The issue of who elects the adjudicator of the game (game-master) is still open, and we shall not discuss that further here.

15.2. CLASSIFICATION OF GAMES

In the proof above, we made use of a centralized construction. This is not strictly necessary for the currency. It is sufficient that every agent or player participating in the game should share a common knowledge of the valuation standard for currency and obeys a constant sum rule. The knowledge of the currency standard can spread epidemically through a promise graph to achieve the same result, provided the agents all abide by the same rules. This requires a special source node for the standard, which can be identified as A, so there is still a adjudicator node. The nodes must subordinate themselves to a set of rules that preserves:

$$\sum_{\pi_i,i,j} v_i(A_i \xrightarrow{\pi_i} A_j) = \text{const.} \tag{15.8}$$

This cannot be done without relinquishing part of their autonomy. Turning the argument around, it is likely that constant sum games do not represent a realistic interaction of autonomous agents.

15.2.6 MULTIPLE STRATEGIES AND MIXTURES

We have tacitly identified the promise of services with strategies in the interpretation of a game in normal form. A promise becomes an alternative in the expectation of rational play. However, there are subtleties about actions and promises to be addressed. What if, for instance, two promises are mutually exclusive? What happens when we form mixed strategies? There are two ways we can envisage mapping games onto collections of promises:

- Strategies are binary: either we keep a promise or do not keep a promise.

- Strategies are multiple: distinct alternative promises for future action, some of which might be mutually exclusive.

Consider the game matrix for a two-agent game below, with payoffs..

(1,2)	π_1	π_2	π_3
π_4	(10,5)	(4,6)	(3,2)
π_5	(0,5)	(7,2)	(5,4)
π_6	(10,5)	(6,6)	(8,3)

The normal solution concept for such a game is to search for an equilibrium configuration of strategies for the players. Generally, such equilibria required mixed strategies to balance their books. However, this leads to a difficulty in interpretation for promises. Some promises might be mutually exclusive, some promises might break other promises, etc. Thus we cannot simply search for a numerical equilibrium for this matrix without applying the logic of promises to the problem also. The formal ambiguity can be dealt with by making promises conditional on certain prerequisites. However, there will be many games represented in the literature that assume information channels that are unavailable to autonomous agents. Readers should therefore take care to remember the implications of complete autonomy. In particular, two agents can only be assumed to share common knowledge about π if they agree[BFa]:

$$A_1 \xrightarrow{\pi} A_2$$
$$A_2 \xrightarrow{U(\pi)} A_1. \qquad (15.9)$$

The issue of mixed strategies is something that must be addressed in both of the cases above. In a zeroth-level approximation to Promise Theory, one generally makes a default assumption that promises are always kept. This allows one to discuss the logic of promise graphs. However, as soon as one considers the matter of observation and repeated trials, and takes seriously the issue of why agents would keep promises, this default assumption becomes too simplistic to model real situations. We discuss some of these matters in the next section. For the remainder of the paper, we restrict out discussions to two-person games (or games that are two person factorizable).

15.3 REPEATED BARGAINING: CONDITIONAL PROMISES

A dilemma that is faced by agents is whether to favour a long term benefit with a certain risk, or short-term guaranteed rewards by choosing a policy towards neighbours. If an agent only interacts once in its lifetime with another, it has nothing to lose by betraying the trust of the other. However, if there is the possibility of a future reprisal against it, the reckoning is a different one[Nas96].

Agents have the possibility of *trading* promises of different types using a valuation based on an agreed measure of common currency, as pointed out in ref. [BFa]. This allows them to secure long term relationships with one another in a way that secures stability of interaction. Selfish agents will only predictably continue to interact if they receive some net payoff from the interaction.

However, in a real world pervasive computing scenario, one must be careful in attributing the failure to comply with promises to non-cooperative behaviour, as one might in economics. There are several reasons why an agent might fail to live up to its commitments in a cooperative relationship:

- Environmental uncertainties beyond the control of the agent intervene (noise).

- Unwillingness to cooperate emerges (change of policy).

15.3. REPEATED BARGAINING: CONDITIONAL PROMISES

- The failure of a dependency agreement, which invalidates a conditional promise (a third party spoils it).

This is not an extensive list, but each case is significant and important to the fragility of developing a regional policy in the ad hoc assembly of individuals.

So, given short term certainty versus long term risk, how might agents choose to behave? Suppose they surmise an incentive to cooperate:

1. Under what conditions would cooperation emerge and be stable in a group of selfish-minded individuals?

2. What kind of *behaviour* is it most beneficial for each individual to choose (strategy)? i.e. How does one harmonize selfish interest with common goals?

3. What might be done to promote the emergence of cooperation in a group by altering their policy for interaction?

The preferred solution method of such game theoretical problems is the concept of Nash equilibrium[Mye91]. An equilibrium is the end result of a 'tug of war' between players. Game Theory therefore embodies a notion of *stability* in the result of the contest. This is an appropriate model for management. The restriction to two players simply means that, in a graph of interactions between individual agents, each negotiation is made between individual pairs, without reference to third parties. This is also what is appropriate for an autonomous environment. We recall some basic notions from Game Theory.

Definition 83 (Two person iterated game). *A game consist of two players or agents, which interact and make choices. For each choice i, by player a, and counter choice j by the other player, they each receive a payoff, which is determined by the payoff matrix* Π_a *with components* $[\Pi_a]_{ij}$ *(see fig. 15.6). The game can consist of one or several rounds. A game of several rounds is called an* iterated game.

This matrix form of the game represents the benefits and losses (hence risks) during *one* encounter between two agents. If they should meet again, and the previous encounter might have influence on how the agents act next time they meet, this is modelled by several *iterations* of the game. Each iteration hence represents one interaction between two agents, and this interaction could be defined in several ways.

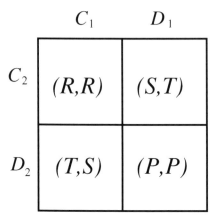

Figure 15.6: The payoff matrix $\Pi_{(1,2)}$ representing rewards for a single move in a dilemma game. The components $[\Pi]_{ij}$ represent the four quadrants for $i,j = 1, 2$

Definition 84 (Move and strategy). *The choice of action $\vec{M}_{a,i}^{\mathrm{T}} \in \{(1,0),(0,1)\} = \{\vec{C}, \vec{D}\}$ made by agent a in round i of a game is called a* move. *These concern keeping or breaking a given promise to the other player. For historical reasons, this is referred to as 'cooperation' (C) and 'defection' (D), see fig. 15.6. A complete sequence of choices $\{\vec{M}_{a,i} | \forall i\}$ for agent a takes over the entire game, constitutes the* strategy *of that player.*

Moves are in fact micro-strategies in the interpretation of the dilemma game as a normal-form, strategic game with constant payoffs. To avoid confusing micro- and macro-strategies, or extensive and normal form games, we use these terms: move and strategy.

To account for the effects of a receding history, the total payoff of a player in an iterated game is generally a discounted sum of the payoffs in the individual rounds[Axe97, Axe84]. Let $M_{a,r}$ be the move made by agent a in the rth round of the game, and let Π_a be the payoff matrix in this round for agent (player) a, given a counter-player $\bar{a}$, then over m moves,

$$\Pi_{\text{total},a} = \sum_{r=1}^{m} \delta^m \, (\vec{M}_{a,r}^{\mathrm{T}} \Pi_a \vec{M}_{\bar{a},r}). \tag{15.10}$$

The factor $\delta \leq 1$ is called the 'discount parameter'. One normally imagines oneself fixed at the present and looking out into (i.e. predicting) the future, based on past experience, so that m recedes into the future. What happens in the distant future (or distant past) is less important than what is immediately upon us, so its importance to the payoff is reduced in weight, or *discounted*. This duality between past and future is the same as that in all probabilistic theories in which evidence from the past is used as a model for the future.

The dilemma game has proven to be a very powerful tool for describing cooperation. The goal of this method is typically to find stable *solutions* to the iterated game, by balancing short and long term gain against one another. The iterated game is played for a given number of rounds and the most profitable player, either during one round, or in the long run (several iterations) wins. The actual winner of a game is not really of interest to us in policy based management. What is important is the average equilibrium outcome, i.e. which strategies or choices of promises agents choose to keep. To be able to predict the outcome of such an iterated game, one needs:

- The strategies of the other players involved in the game
- The number of encounters/interactions with the other players
- The payoff/reward for the different combinations of encounters
- The start state.

This translates into finding out what kind of behaviour is most beneficial for an agent in the long run, dependent on what kind of environment the agent acts in, how often each agent meets agents with certain behaviour and so on.

15.4 PROMISES: CONSTRUCTING POLICY

It is plausible that there is a connection between the two person games studied in the theory of cooperation and the likelihood that promises are kept between individuals in a pervasive computing environment. We shall now show this.

A promise graph $\Gamma = \langle I, L \rangle$ is a set of nodes I and a set of labelled edges or links between them L. Each directed edge represents a promise from one agent to another. The set of links L falls into several non-overlapping types, representing different kinds of promise. Thus each promise begins with a label π_i, denoting its type, and with a specification of the promise constraint, which it promises to uphold.

Lemma 12 (Games are bilateral promises). *Let a two person strategic game be represented by a pair of 2×2 matrices Π_a, and moves $\vec{M}_i$ where $A_i, i = 1, 2$ are the agents and $a, b = 1, 2$ label the choices to cooperate or defect in the payoff matrix, e.g. $[\Pi_i]_{ab}$. There is a mapping of each game between the players onto a pair of promises.*

Proof. For the players in the game, introduce a promise of type π_a and a return promise of type π_b, which are traded in the manner of fig. 15.7a. Now let $\vec{T}(\cdot)$ be a 'truth' function that maps any promise into a one of the values $\{\vec{C}, \vec{D}\}$, indicating whether the promise is obeyed or not by the promising agent. The truth function may be associated with a move in a game, of a type that maps one to one onto the promise type, and hence may be written equivalently:

$$\vec{T}(A_1 \xrightarrow{\pi_i} A_2) \rightarrow \vec{M}_1 \quad (15.11)$$

$$\vec{T}(A_2 \xrightarrow{\pi_j} A_1) \rightarrow \vec{M}_2 \quad (15.12)$$

$$\left[v_a\left(A_1 \xrightarrow{\pi_i} A_2, A_2 \xrightarrow{\pi_j} A_1\right)\right]_{ab} \rightarrow [U_i]_{ab} \quad (15.13)$$

where v_i is the matrix-valued valuation function for agent i, which interprets the value of the mutual relationship to agent i; π_a represents the promise body, which is labelled by the types a, b. We must have either $(A_1, A_2) = (A_1, A_2)$ or $(A_1, A_2) = (A_3, A_3)$ to mediate the game. The proof is now trivial noting that the arrow is from promises *onto* games, by a policy determined function v_i. □

Note that, although we can identify a game with a pair of promises, not all promises can be meaningfully viewed as games, as we see in section 15.6. Any reasonable function v_i can be used to defined the mapping from promise to game, since the function v_i is to provide a autonomous *valuation* of received promises by each agent A_i. The game can then be used as an estimator for the rational behaviour of the agents as they interact.

A consequence of this relationship between promises and games is that promises must also specify a policy for stabilizing the cooperative relationship over time, i.e. a strategy for the entire iterated game.

15.5 MANAGEMENT DILEMMAS

A common approach for modelling and analyzing cooperation is to use dilemma games from two-person Game Theory[Axe97, Axe84]. Here two individuals, without strictly opposing interests, have the possibility of exploiting each other's willingness to cooperate for selfish gain. This is a scenario that has shown to fit several real-life cases[Axe97, Axe84]. It is known as a cooperative dilemma.

In the dilemma game, each player can choose between two different moves, *cooperate* (keep promise) or *defect* (i.e. do not keep promise). The interpretation of the dilemma game symbols is thus: T is the payoff for receiving services from another agent, while refusing to contribute; R is the reward two agents receive when both are cooperating; P is the inconvenience both agents experience when they both refuse to collaborate; and S is the inconvenience of servicing an agent that refuses to contribute anything back.

15.5. MANAGEMENT DILEMMAS

There are many possible outcomes, which depend on the relative sizes of these parameters. For the game model to qualify as a Prisoner's Dilemma, in the common usage of the term, certain requirements need to be fulfilled: (i) $T > R > P > S$, and (ii) $R > \frac{(T+S)}{2}$. These requirements ensure that conclusions associated with the game are true.

A player can win the maximum payoff by 'defecting' (failing to act according to its own promises) when the other player is 'cooperating' (acting according to its promises), but if both players are defecting, they both receive a lower score than if they both had cooperated. Constraint (ii) assures that mutual cooperation is more beneficial than alternating cooperation and defection between the two players.

What makes this game model interesting, is the controversy that even though the choice *defect* is *dominant*, both players will actually end up with a better result cooperating in the long term. This model is completely analogous to a scenario of autonomous agents, looking to establish agreements (promises) between one another, in order to create a larger system voluntarily. There is a potential long-term reward associated with providing services for other peer agents, but avoiding a commitment could thus seem beneficial in the short run.

We model this situation with multi-move games in which one agent can follow the history of its interaction with another and *punish* previous defections from the other player by failing to behave cooperatively in return. An example strategy, which has received much attention in the literature, is *tit for tat*. This has proven to be very successful in playing cooperative dilemmas with a good average performance. Tit for tat is a macro-strategy and it is defined as follows:

Definition 85 (Tit for tat). *Let r label a sequence of rounds in a game, and let $\vec{M}_{1,r} \in \{\vec{C}, \vec{D}\}$ be the move taken by agent 1 in round r, etc. The tit for tat strategy is defined by:*

$$\vec{M}_{1,1} = \vec{C} \qquad (15.14)$$

$$\left.\begin{array}{r}\vec{M}_{1,r} = \vec{M}_{2,r-1} \\ \vec{M}_{2,r} = \vec{M}_{1,r-1}\end{array}\right\} \quad r > 1. \qquad (15.15)$$

Notice that the agents always start out cooperating, or obeying policy rules, and then do against the opponent what the opponent did on the previous move.

The success of this strategy, in dilemma games, is characterized by the fact that it gets high scores on average, even though it does not necessarily always win (i.e. it does not always maximize the utility of its interaction). This strategy has several beneficial qualities: it is simple, it rewards cooperation, it punishes defection and it is forgiving, hence single defection noise (including accidents) does not harm a player too much. The disadvantage is that it does not exploit recovery from random noise well enough. It tends to end in fruitless reactive oscillations between agents, once they have started. A continuous generalization of the game has further instability problems[BBJF]. Some thing is required to dampen out the oscillations caused by random errors. A small amount of noise to unleash oscillations throughout a peer network, in which the agents had dependent promises.

The real dilemma in these games is this: if an agent in a trading relationship does not retaliate to broken promises, it will allow itself to be exploited, exhausted and stability will be compromised. However, if it does retaliate, it participates in the contributing to a policy instability in the network.

15.6 EXAMPLES OF GAMES AND THE ECONOMICS OF PROMISES

What is the complete relationship between games and promises? We need to establish how the strategies interact and affect one another. Some relationships are purely fortuitous, with no guarantees of receiving a payoff for their actions. Others interactions are games which are bound by a trade of mutually beneficial services. Thus not all promises are games, but all dilemma or bargaining games involve promises. How does the probability of keeping a promise relate to the experience gained from iterative dilemma games, and the discount parameter?

Since the cooperative dilemma gives us a natural notion of stable equilibrium for autonomous agents, let us ask: in what sense is a promise a cooperative dilemma, and which agents fall outside this model?

Consider two nodes, node A and node B, which each represent an autonomous node. As described in [BFa], a promise $a \xrightarrow{\pi} b$ from a to b is represented in graph notation by a directed edge (arrow) from node a to node b, and the label on the edge defines the *content* of the promise.

Definition 86 (Policy dilemma). *A bi-lateral promise between two nodes, in which cooperation is interpreted as* keeping *the promise, and defection is defined as not keeping the promise. In order to qualify as a bargaining game, both nodes must receive some utility from the bilateral arrangement.*

15.6. EXAMPLES OF GAMES AND THE ECONOMICS OF PROMISES

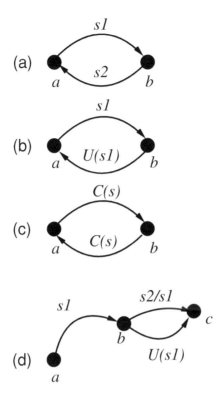

Figure 15.7: Some 2 person games and their corresponding promises.

Consider the images in fig. 15.7. Here we see a number of promise graphs, some of which are policy dilemmas and some of which are not. There are three basic categories of promise[Bur05]:

1. In a service promise of the form $a \xrightarrow{s} b$, node b is the recipient of something of value (the promised service).

2. In a use-promise, $a \xrightarrow{U(s)} b$, node a promises node b that is will avail itself of a service s, which is has been promised by some node. In this case, node b does not receive anything tangible, except an assurance that a will use the service it has been offered.

3. Similarly, in a cooperative promise $a \xrightarrow{C(s)} b$, node a promises node b that it will imitate b's behaviour with respect to service s. Again, the only thing b receives is an assurance.

We must now ask: are all the graphs in figure 15.7 dilemma games? In case (a) the two nodes exchange services or other similar promises of some value to one another. We may consider this interaction as representing a bargaining game between the players, if the promises are judged beneficial to the agents, e.g. BGP peering agreements between service providers.

In case (b) however, the assurance of usage does not seem enough to bargain with. Indeed, node a says "if you do not give me s, I will not use it!". This does not seem like a strong bargaining position. One could similarly apply (b) to a cooperative promise: "If you do not give me s, I will not do as you do with regard to s" (give myself s). Again, there are no grounds for bargaining. This could be misleading however. Consider the example of refuse recycling or garbage collection in society. Users promise to deliver their garbage and the collectors agree to take it. Both parties can benefit from this because by *not* cooperating they would incur a loss. This warns that caution is required in jumping to conclusions.

In case (c), there is a game, since the two promises are on equal footing: "if you don't do as I do, I won't do as you do". In both cases, the promises are of equal value and so they have a symmetry of mutual assurance, and we can define them to be a game. Indeed, case (c) can easily be shown to be a stable fixed point.

Case (d) contains no game either. Node c gives nothing in return to node b for its service s_2; node b offers nothing in return to node a for s_1.

The services are promised without any trade taking place. In order to establish a game, an agent must be able to exert an influence against its peers. It must have something to bargain with. There are two kinds of nodes that do not fit into this contract of stability, easily located by network analysis.

Definition 87 (Opportunist sink). *A network sink (a node with out-degree zero) is opportunist. It is vulnerable to loss of service, as it offers nothing in return. It is a non-profit node.*

Definition 88 (Altruist source). *A network source (a node with in-degree zero) is an altruist. It is vulnerable and exploited, as it gives without taking anything in return. It is a 'single point of failure' for the dependent promises.*

Thus network analysis could enable us to repair the lack of economic incentive to cooperate in case (d) by modifying the graph to allow for some chain of payment (see fig. 15.8).

Thus longevity of relationships provides a direct motivation for introducing a trading currency.

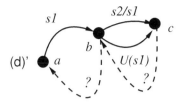

Figure 15.8: An act of uncertain altruism can be turned into an incentivized bargaining game by assuring some sustaining payoff to the altruist.

15.7 MINIMUM INCENTIVE REQUIREMENTS

We can now propose some minimal requirements for sustainable promises, based in economic incentives.

Proposition 1 (Common currency). *The need to bargain or trade with different currencies motivates an undirected graph of common currency, representing bilateral promises.*

Proposition 2 (Principle of sufficient remuneration). *Every agent should receive a 'handshake' promise in return for giving a promise. The handshake can be valued in any currency used by the agent.*

This proposition can easily be proven if one posits equilibrium as a criterion for a well-formed policy. We shall not attempt to squeeze this into the present work.

The idea of offering different interpretations of currency, by which to motivate incentive in promise-keeping, has been considered also Piaget[Pia95]. In his view, societies of agents bind into working constellations because they value one another by different criteria. Consider a transaction performed between two 'agents', represented by A and B, where A performs some action *on behalf* of B. Piaget assumes the following basic conditions, which differ from our starting point:

- A common scale of values for all agents.
- Conservation of values in time.

and remuneration values: i) Renouncement value r_A, representing the *investment* made by agent A in order to perform an action on behalf of B. ii) Satisfaction value s_B, representing the satisfaction expressed by B as a result of receiving the action from A. iii) Acknowledgment value t_B: representing the acknowledgment made by B to A as a result of receiving the action. iv) Reward value v_A, representing the associated value of receiving the acknowledgment from B, as experienced by A. Piaget distinguishes between so-called *real* and *virtual* values, where r and s are the real values, and t and v the virtual values in the transaction. The virtual values turn into real values when they are 'cashed in'.

Many writers assume the imposition of law and obligations forms a sufficient basis for civil order, without incentive; however, this assumption can't be supported without a considerable foundational infrastructure involving promises and voluntary cooperation.

Example 150 (Legally binding agreements). *The literature of games and economics often refers to 'binding agreements' in the sense of agreements that are enforced by some legal framework[BG02]. In Promise Theory there can be no such outside legal framework, unless one builds it from promises. Such a framework is necessarily complex. A third party (government or law court) must promise to make laws known. All agents must promise to use the laws. They must further promise to make visible any contracts and practices they make to an arbitrator, who in turn must promise to inform law-enforcers of any breakage of laws. Enforcers in turn could enforce an isolation (incarceration) of an agent.*

To make an agreement binding by force, one must subordinate oneself to an outside force, or suffer large consequences. Such a scenario takes considerable infra-structure to achieve, which shows that commonly held assumptions about civil society often represent very complex scenarios (which is perhaps why criminals stand a chance of succeeding in breaking laws). Ironically the main reason for the contractual success of such a civil arrangement is a sound economic basis for agent outcomes, rather than the threat of punitive action post hoc.

CHAPTER 16

SEMANTIC SPACETIME

A collection of agents effectively acts as a set of locations, which can promise properties such as occupancy, colour, and adjacency. The promise bindings between them act as links in a directed graph. These are the basic elements of what we understand as 'space'. As properties change, the change of states forms a clock that ticks. This is what we understand as 'time'. The observation that we can form a graph-like spacetime as a promise theoretic model offers a way to inject functional semantics into a model of things. We call this model *semantic spacetime*[74].

16.1 EXTENDED AGENT STRUCTURE

The ability to exchange information may be regarded as synonymous with the concept of adjacency, i.e. 'being next to', for all intents and purposes. We shall take this view here. The common conception of space is that of an ability to observe and transport something (information) from observer to observer. This might seem peculiar from the viewpoint of an absolute Newtonian spacetime, but the view is quite compatible with an Einsteinian view. The idea of 'distance', for example, is just one of many possible associations between agents that change with perception, circumstance, and individual capability. Thus, to describe spacetime as a collection of agents, we must begin by rekindling the notion of distance from the more primitive concept of adjacency of agents, i.e. locations.

Adjacency can be promised by any agent to any other, by promising any information to promisee. This assumption is totally unspecific. Indeed, all promises have this structure, thus we may infer that *any* promise binding implies a typed notion of adjacency between agents. Agents are next to one another if they promise to be. The conventional notion of points being next to one another in space as just one particular type of relationship between points (agents). Any other promise would work in the same way.

Being close to, being able to see or hear a neighbour, being able to point to, or even being attracted to, are examples of adjacency, as they name a specific target. Thus an adjacency promise is more than mere continuity[75].

Definition 89 (Adjacency promise). *An adjacency promise is a promise that relates an agent x_i to another specific unique agent x_j ($i \neq j$), and may give a local interpretation to a relative orientation i.e. direction between the two.*

Agents making an adjacency promises to more than one agent cannot be exchanged without changing the linkage. Thus adjacency is a form of *order* (see section 16.3).

The scope of any promise defaults to the two agents involved in the adjacency relationship, but it could also extend beyond them, allowing others to observe the relative positioning of points, allowing in turn the coordination of distributed behaviours. In semantic structures like swarms (flocks of birds, or shoals of fish), nearest neighbour observations are sufficient to maintain the coherence of the emergent cohesion, suggesting that a spacetime formed from autonomous agents with nearest neighbour interactions could be sufficiently stable without long range interactions.

16.1.1 Continuity and basis

How do ideas such as continuity arise in a semantic spacetime? Suppose we wanted a concept of travelling North. How can this be understood from an agent perspective? The concept of North-ness is non-local, and uniform over a wide region. In order to image continuing in the same direction, we also need to know about the continuity of directionality. Direction too is thus a non-local concept. When we speak of direction, we mean something that goes beyond who are the agents closest to us.

Any agent can promise to bind to a certain number of other neighbours, calling its adjacencies to them with the same name (say North, South, etc), but why would the next agent continue this behaviour? How does each agent calibrate these in a standard way?

Definition 90 (Spatial continuity). *If a promised direction exists at a certain location x_i, it continues to exist in the local neighbourhood around it.*

Directionality is a property like a colour, as discussed in section 4.5.8, using the matroid or basis set pattern. Membership in a basis set is a semantic convention used by observers. It cannot be imposed. A point in a space need not promise its role in a coordinate basis, because that information is only meaningful to a godlike observer of all agents, and could simply be ignored by the observer. An agent can promise to be

16.1. EXTENDED AGENT STRUCTURE

adjacent to another agent, but to propose its own classification as a member of some basis would be to impose information onto others from a different viewpoint. By autonomy, each agent is free to classify another agent as a member of an independent set within a matroid that spans the world it can observe.

The dimensionality of spacetime perceived by every agent is the the rank of the matroid it chooses to apply to the agents it can observe as neighbours. The consequence of this is that spacetime can have any dimension that is compatible with the adjacencies of the observer. The notion of dimensionality experienced by the elements of a promised space is different for every agent, at every point.

Law 14 (Matroids are local observables). *Every autonomous agent decides its own set of independent sets to span a space. Hence direction is a local observer view.*

Consider an ordered sequence of agents x_i that are mutually adjacent. An agent x_i recognizes a direction μ if it promises adjacency ($+\text{adj}_\mu$) along a locally understood direction μ to a subsequent neighbour x_{i+1}, *and it promises to accept adjacency* ($-\text{adj}_\mu$) with a previous neighbour x_{i-1}:

$$x_i \xrightarrow{+\text{adj}_\mu} x_{i+1} \quad (16.1)$$

$$x_i \xrightarrow{-\text{adj}_\mu} x_{i-1} \quad (16.2)$$

$$x_i \xrightarrow{+\text{adj}_\mu} x_{i-1} \quad (16.3)$$

$$x_i \xrightarrow{-\text{adj}_\mu} x_{i+1} \quad (16.4)$$

for all x_i. Or in shorthand:

$$x_i \xrightarrow{\pm\text{adj}_\mu} x_{i\pm1}, \forall x_i. \quad (16.5)$$

We shall need to say what happens at edges where we run out of x_i (see section 16.2.7). These promises are local but require long range homogeneity between the agents, i.e. the condition $\forall x_i$ is a non-local constraint. It is equivalent to promises by every agent to conform to these promises:

$$x_i \xrightarrow{C(\text{adj}_\mu)} x_j, \forall i,j \quad (16.6)$$

The issue is not the numbering $i = 1\ldots N$ of the agents, as this may be freely redefined. Any local agent will bind to another exclusively and the ordering can easily emerge by self-organization, however, the notion that all of the agents or spacetime points would coordinate with long range order in keeping these promises does beg an explanation.

A multi-dimensional interpretation of the different spanning sets, does not really add further difficulties, but emphasizes further the non-local cooperation in terms of promise homogeneity. If we choose a 3 dimensional basis with coordinate names (x, y, z),

$$(x_i, y_j, z_k) \xrightarrow{\pm \mathrm{adj}_\mu} (x_{i \pm 1}, y_j, z_k) \tag{16.7}$$

$$(x_i, y_j, z_k) \xrightarrow{\pm \mathrm{adj}_\mu} (x_i, y_{j \pm 1}, z_k) \tag{16.8}$$

$$(x_i, y_j, z_k) \xrightarrow{\pm \mathrm{adj}_\mu} (x_i, y_j, z_{k \pm 1}) \tag{16.9}$$

The tuples are promises that belong to each local agent, whereas the bindings are the result of an interaction between the agents. The directional names belong to the local agent's coordinate basis, and the non-local cooperation is assumed homogeneous. This leads to the possibility of a collection of misaligned, non-oriented agents self-organizing into a crystal lattice. In this way, a space can acquire *long range order* with only local, autonomous promises, provided they are homogeneous over a sufficient region.

16.1.2 Scalar promises—material properties

A question that becomes relevant when we approach the matter of spacetime semantics is whether there is a basic difference between empty space and something material that fills it? Are the agents of 'space' fundamentally different from the agents of 'matter', or merely different states of the same substrate?

Example 151 (Data memory). *Imagine a blank storage array, which becomes filled with data. Is the absence of an intended content fundamentally different from the promise of an unspecified value? Is a tissue of stem cells different from cells that have been given material semantics by expressing differentiated types? Both have DNA; they merely express different promises. Is empty space merely an agent without a promise to behave like matter? There is at least the possibility that matter is simply the breakdown of indistinguishability in space.*

Definition 91 (Scalar promise). *A promise made without reference to any other promise made by another agent.*

Scalar promises imbue elements of spacetime with intrinsic properties. Conditional scalar promises are also possible, that rely on a promise from a different agent. The matroid construction (see section 4.5.8) for calibrating scalar properties is an example of this. This blurring between scalar and vector is the essentially the trick used in theoretical physics for introducing scalar properties by extra 'hidden dimensions'.

16.1.3 VECTOR PROMISES AND QUASI-TRANSITIVITY

We can generalize promises to vector and even tensor relationships for agents that refer to one or more other agents. Vectors and tensors describe relationships between locations.

Definition 92 (Vector and tensor promise). *Promise made in reference, relative to the promises made by other agents, with directions defined by their bindings.*

Vector promises are, in principle, interpretable as one of the following cases: i) A_1 can influence A_2 (causation), ii) A_1 is connected to A_2 (topology), iii) A_1 is part of A_2 (containment) e.g.

$$A_1 \xrightarrow{\text{Causes}} A_2 \tag{16.10}$$

$$A_1 \xrightarrow{\text{Precedes/Follows}} A_2 \tag{16.11}$$

$$A_1 \xrightarrow{\text{Affects}} A_2 \tag{16.12}$$

$$A_1 \xrightarrow{\text{Is a special case of}} A_2 \tag{16.13}$$

$$A_1 \xrightarrow{\text{Generalizes}} A_2 \tag{16.14}$$

Adjacency too may be considered only quasi-transitive, for while A next to B next to C does not imply that A is next to C, if we reinterpret adjacency as connectivity, we can make transitivity true as long as all promises are reciprocal and accepted, e.g. π_L: A is to the left of B is to the left of C.

$$\pi_{\text{adj}} \rightarrow \begin{cases} \pi_{\text{connected}} \\ \pi_{L,R} \\ \pi_{N,S,E,W} \\ \pi_{\pm\mu} \end{cases} \tag{16.15}$$

Clearly if A causes B and B causes C, there is a sense in which one might (at least in some circumstances) interpret that A causes C, hence there is a kind of transitivity. In mathematics one usually introduces a generator of a translational symmetry which can be repeated more than once to bring about a sense of continuity of motion—this is an imposition, not a local promise.

Next there are container models, analogous to bi-graphical positioning:

$$A_1 \xrightarrow{\text{Is contained by}} A_2 \tag{16.16}$$

$$A_1 \xrightarrow{\text{Is found within}} A_2 \tag{16.17}$$

$$A_1 \xrightarrow{\text{Is part of}} A_2 \tag{16.18}$$

$$A_1 \xrightarrow{\text{Is eaten by}} A_2 \tag{16.19}$$

These generate forest graph relations. Thus we have a way of incorporating both types of spatial semantics in the promise framework, and we can link translation and containment through their quasi-transitive nature. These promise types provide a notion of *spatial continuity*.

Promises that cannot be made into a succession of symmetrical translations belong to the singular scalar promises. They represent promises about self, rather than about relationships to others (in functional terms they have arity 1). Such expressions may be formulated and interpreted in two ways, depending on who or what are the recipients of the relationship: as promises, or as general associations between topics in a topic map.

16.2 ADJACENCY BETWEEN AGENTS

In order to establish which agents are next to one another, in a promise graph, we need to build a picture of mutual interaction. In Promise Theory, it is not automatically true that, if A_1 is next to A_2 then A_2 is also necessarily next to A_1. Spacetime can have one-way streets and mirrors. This is the price one pays for properly defining the semantics of locality.

16.2.1 LOCALITY

By now, it should be clear that there is a difference between what may be perceived by a number of agents promising ordered relationships amongst themselves, and what a 'third party' observer can discriminate about the situation from a remote location. We need to explore this more formally. Let's explore the meaning of $A_1 \preceq A_2$ in an autonomous sense. In semantic spacetime, agents undergo connections to neighbours, which are the only locations they have direct contact with or 'knowledge of'.

In principle the exchange of any kind of message could be seen as evidence of proximity. We can take this a definition of adjacency: the ability to exchange a promise with a neighbour[76]. In an autonomous world, this is not a mutual promise. Let's restate the adjacency slightly as follows:

Definition 93 (Adjacency promise (alternative)). *An adjacency promise is a promise made by an agent A_i to another specific unique agent A_j ($i \neq j$), about its claim to a local interpretation to a relative orientation i.e. direction between the two. We write:*

$$A \xrightarrow{+\text{adj}} A', \qquad (16.20)$$

i.e. I promise that I am next to you, from my perspective.

The agent may, in addition, distinguish different kinds of adjacency ($x, y, z, \ldots$), but we can drop this for now. Agents making an adjacency promises to more than one

16.2. ADJACENCY BETWEEN AGENTS

agent cannot simply be exchanged for one another without changing the linkage. Thus adjacency is the beginning of a form of local *order*, which may lead to a *long range order*, by mutual cooperation[77]. Let us now summarize how many primitive promises are needed to bind adjacent points in a spacetime.

Definition 94 (Adjacency promise binding). *A bundle of bilateral promises, analogous to a contract, binds an agent A_n with another agent A_{n+1}, promising a channel between them.*

- A_n *promises that A_{n+1} may transmit (+) directed influence to it.*
- A_{n+1} *promises to use (-) A_n's offer.*
- A_{n+1} *promises that A_n may transmit (+) directed influence to it.*
- A_n *promises to use A_{n+1}'s offer (-).*

This can be interpreted as adjacency. In a forward (retarded) direction,

$$A_n \xrightarrow{+\text{adj}} A_{n+1} \tag{16.21}$$

$$A_n \xrightarrow{-\text{adj}} A_{n+1} \tag{16.22}$$

$$\tag{16.23}$$

And in the reverse (advanced) direction:

$$A_{n+1} \xrightarrow{+\text{adj}} A_n \tag{16.24}$$

$$A_{n+1} \xrightarrow{-\text{adj}} A_n \tag{16.25}$$

It feels strange according to the traditions of physics to think of being next to something as a promise, locally decidable by every location individually, but this is not as strange as it feels. An event horizon is a 'one way glass', for instance. In computer science, the adjacency of locations is a virtual decision made by processes to interact.

16.2.2 NON-LOCAL EXTENDED CAUSAL ORDER

Building extended lattices, like crystals or vector spaces, is not as trivial as it appears once we strip away the promises we take for granted about location. No agent is in a position to know its role in an extended region. Suppose an agent tried to promise an ordered relationship with its neighbour 'I am greater than you'. On what basis could it make this assessment. Each agent is at liberty to make the same claim, so that

$$A \xrightarrow{+\text{greater than}} A' \tag{16.26}$$

could be countered with

$$A' \xrightarrow{+\text{greater than}} A. \tag{16.27}$$

This stalemate could only be broken by a third party observer O with access to all of the agents. But this is not so straightforward either: now local names are no longer sufficient for O to keep track of the agents. O has no information about A_i unless the nodes are *distinguishable*:

$$A_i \xrightarrow{+\text{name}=A_i} O \quad (A_i \neq A_j) \tag{16.28}$$

$$O \xrightarrow{-\text{name}=A_i} A_i \tag{16.29}$$

$$A_i \xrightarrow{+A_i > A_k} O \tag{16.30}$$

$$O \xrightarrow{+A_i > A_k} A_i \quad (A_i \neq A_k) \tag{16.31}$$

$$\tag{16.32}$$

From this information, O is not able to promise an opinion, conditionally on the basis of what it has been promised first hand:

$$O \xrightarrow{+(A_i > A_j < A_k \ldots) \mid (A_i > A_m),(A_j > A_p),(A_k > A_q)} A_?. \tag{16.33}$$

Whether the agents A_i also make these order promises or impositions to one another is irrelevant to O, because (we assume, in the first approximation that) it only receives what each of the agents individually promises from them. If each agent acts unilaterally:

$$A_i \xrightarrow{+\text{greater than you}} \blacksquare \quad A_j \quad (A_i \neq A_j) \tag{16.34}$$

or if agents act by invitation (with prior affinity)

$$A_i \xrightarrow{+\text{greater than you}} A_j \quad (A_i \neq A_j) \tag{16.35}$$

$$A_j \xrightarrow{-\text{greater than you}} A_i \quad (A_i \neq A_j) \tag{16.36}$$

$$A_j \xrightarrow{+\text{less than you}} A_i \quad (A_i \neq A_j) \tag{16.37}$$

$$A_i \xrightarrow{-\text{less than you}} A_j \quad (A_i \neq A_j) \tag{16.38}$$

In order for this to work, there would need to be a common language understood through the interactions that matches "I precede you" with "I follow you" for each agent: a *gradient* promise.

If each agent also passed on overlapping information about its neighbours, then suddenly topology becomes a second order process—a statistical assessment made by every observer on a timescale greater than the timescale of adjacency interactions between the local agents (points). This implies that any observer capable of measuring

16.2. ADJACENCY BETWEEN AGENTS

relative positions is an agent equipped with sufficient interior memory as to form such a consistent map of assessments. By promising its name (some distinguishable property), we ensure that the observer O can tell the difference between the promises it receives, without imagining a fictitious non-local coordinate system.

We see that our tacit assumption of a global structure is far from easy to arrange by local properties alone. An agent capable to making observations compatible with a coordinate map must be of sufficient interior dimension to possess local memory accumulated from agents it is in contact with my messaging channels. Thus, any view of spacetime as a manifold is fundamentally non-local, from an information theoretic perspective.

16.2.3 ORDERED AGENTS

Suppose now that a collection of agents effectively (and emergently) promises to crystallize into a coordinate lattice (figure 16.1), by each promising its role of being next to a neighbour. Now, we need basic distinguishability of agents by their neighbours, in order to specify their precedence:

$$A_1 \xrightarrow{X_1} A_2 \xrightarrow{X_2|X_1} A_3 \ldots \tag{16.39}$$

This is a Markov chain. Filling out the promises in full[Bur14],

$$A_1 \underset{\xleftarrow{-X_1}}{\overset{+X_1}{\xrightarrow{\hspace{1cm}}}} A_2 \underset{\xleftarrow{-X_2}}{\overset{\overset{+X_2|X_1}{\xrightarrow{\hspace{1cm}}}}{\underset{-X_1}{\xrightarrow{\hspace{1cm}}}}} A_3 \underset{\xleftarrow{-X_3}}{\overset{\overset{+X_3|X_2}{\xrightarrow{\hspace{1cm}}}}{\underset{-X_2}{\xrightarrow{\hspace{1cm}}}}} A_4 \ldots A_n \tag{16.40}$$

This promise structure leads to serial order, at the expense of non-local knowledge of neighbouring agents. Notice the apparently acausal feed-forward promises at each that complete the logical semantics 'I have made a conditional promise, and I promise that I have the condition in hand'.

The notation in (16.40) suggests that every agent need to know a unique name of its nearest neighbours. That may not be strictly true, since in a Markov process it suffices to know that one has a predecessor in order to build a ladder. However, if we are interested in maintaining a causal order relation in which each agent knows in which direction a certain vector continues, that information is not clear without labels that go beyond a simple predecessor.

For example, suppose you are a spinning top (which has no intrinsic direction sense) trying to walk in a straight line along across a chess board by hopping from square to square. After each hop, how does the spinning top know which direction it came from in order to continue in the same direction? This clearly has implications for quantities like velocity and momentum along trajectories.

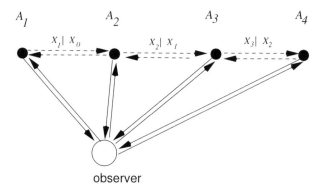

Figure 16.1: A number of agents A_i promise that they are greater than their neighbours and less than the preceding one. For an external observer, there is nothing in the information supplied by the agents that suggests their relative order unless the observer accepts such information from the remote agents on trust. Otherwise it has to trust its own ability to discriminate messages by sensory criteria (e.g. the angle of incidence of the signal on its boundary), which assumes macroscopic interior structure on the observer—suggesting that the size of the observer, and its relation to promise channels, play a role in the ability to assess order.

Again, there is no reversibility yet. For that, we would need an independent chain of promises in the opposite direction:

$$A_1 \xleftarrow{\overline{X_2}|\overline{X_3}} A_2 \xleftarrow{X_3|X_4} A_3 \ldots \qquad (16.41)$$

The property of reversibility is not automatic, if one takes locality as an intrinsic starting point—it has to be promised independently by some or all agents. Why we should observe symmetry between left and right, forwards and backwards, up and down, is unclear. We've come to assume that symmetry is an absence of information, but if we treat all agents as a coordinate lattice, that is not the case. Indeed, the opposite is true. There is long range order.

From equation (16.40), we see that each agent has to be 'aware' of its neighbours in order to form a lattice. This happens because agents are assumed to be autonomous or independent. Normally in mathematics, we assume that points form any structure we want because we impose it thus. In a Promise Theory perspective there is no reason to assume this—and we see no such automatic structure at large scales, so there is no obvious reason to expect it a small scales either.

Example 152 (Crystalline solids). *In crystals, multi-pole distortions of the lattice on the scale of the electron clouds exhibit these non-local effects. The indistinguishability of electrons allows the entire structure to remember some average configuration without*

16.2. ADJACENCY BETWEEN AGENTS

precise and expensive promises to track every electron in the role of a unique messenger M_α. At this scale, the role of agent locations is played by atomic configurations for the different elements, which promise to fall into a finite number of classes from the periodic table.

16.2.4 THE MEANING OF A LINK

The counting of a regular spatial distance is analogous to the counting problem for time. In a metric space, the order and regularity of space is defined by a functional dependence on coordinates. This is often called local, but it is non-local in the sense that the properties over an entire tangent space are presumed regular and coordinatizable, which amounts to a tautology.

Every promise made between agents implies a total ordering of the agents, by virtue of their being unequal. So two agents A_1 and A_n being near or far apart depends on the channel criteria between them: a sequence of autonomously promised orderings. Suppose we introduce a typical partial order relation between agents '$\preceq$', such that:

$$A_1 \preceq A_2 \preceq \ldots \preceq A_n. \tag{16.42}$$

This is only a description of a process built on promises. Each promise is a basic spacetime fabric promise of adjacency, with body adj, as in [Bur14], where an observer can assess each of these promises:

$$\alpha_O(A_1 \xrightarrow{+\text{adj}} A_2)$$
$$\ldots$$
$$\alpha_O(A_{n-1} \xrightarrow{+\text{adj}} A_n) \tag{16.43}$$

The assessment depends on the observer O's self-calibrated understanding of scale. Thus a metric space has no meaning without a local observer view. The observer needs a memory of coordinate associations with known sources to make this picture. Even the pairwise promises in (16.43) above are not sufficient to ensure the total order in (16.42), while it suffices for the individual agents to treat adjacency as a Markov process, for the observer the promises need to be made conditionally because it cannot a priori trust the channels over which information about relative order has arrived.

To say that an agent is close to another requires assessments of $\preceq$ over extended sequences, and potentially different types of promise. To promise distance, we have to accept all the promises of intermediacy. This is a version of the well-known end-to-end problem in service delivery. The observer may be able to promise the distance between A_1 and A_n conditionally, based on the promises from A_n, but this has a high cost:

$$O \xrightarrow{d(A_1,A_n) \mid (A_1 \preceq A_2),(A_2 \preceq A_3)\ldots(A_{n-1} \preceq A_n)} A_? \tag{16.44}$$

In the absence of a god's eye channel, outside of the ordinary adjacency promise linkage, this promise cannot be made without conditional evidence[78].

We can't absolve the problem of basically *trusting* the integrity and authenticity of self-consistent observer experiences when describing spacetime as an independent entity. The implication that information about spacetime locations has to be promised by sender and receiver, and that every receiver forms its own assessment of the information it receives, implies that a coordinate system is a *trusted label system* accumulated over many interactions with different pairs of agents. It is possibly even second hand knowledge passed through intermediaries—a model kept in the memory of an observer to classify subsequent measurements believed to originate from a labelled source. The observer needs to believe it can distinguish the source reliably over multiple observations, i.e. processes need sufficient stability to form a reliable reference system to be called coordinates. This is analogous to Mach's principle about the fixed stars.

16.2.5 RELATIONSHIP TO THE END-TO-END PROBLEM

The existence of the order relation is usually assumed as a basis for structure in mathematical descriptions of space, but it's interesting to see how such a relation can be constructed within the simple rules of Promise Theory. This reveals how there is no automatic reversibility in a fully local theory, contrary to assumption in physics. Imagine

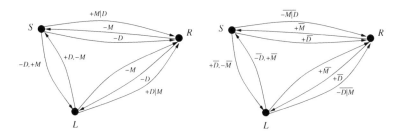

Figure 16.2: The end-to-end promises for a unidirectional process, with complementary transform. The right hand version is a relabelling of the left hand version.

a link L between agents S and R. This takes us back to the end-to-end problem in section 11.4.2. Figure 16.2 shows a situation in which an agent in the role of 'sender' or earlier agent S makes a promise to a receiver or later agent R, with the help of a third party L, an intermediary can be viewed as a 'link' agent. The promised role of L in this example will show what it means for there to be a causal link between S and R. This takes on a special importance when scaling. We can think of the sending of a message by a runner,

16.2. ADJACENCY BETWEEN AGENTS

or the postal order delivery from S to R with the help of a postal service L. The scale is not important, nor the nature of the agents, to the structure of causal message delivery.

For example, the interpretations can be read:

$$+M \leftrightarrow \text{Offer message} \tag{16.45}$$

$$+D \leftrightarrow \text{Offer delivery} \tag{16.46}$$

$$+\overline{M} = -M \leftrightarrow \text{Accept message} \tag{16.47}$$

$$+\overline{D} = -D \leftrightarrow \text{Accept delivery} \tag{16.48}$$

In the figure, we can imagine a distinguishable message M is promised by the source S to a receiver R conditionally, making use of an intermediary L which promises delivery labelled by D. In other words, L will do the actual work, but the intent is signalled by the endpoints. S promises $+M|D$, which reads 'I promise you M if I get D from elsewhere'. The source further promises that it has acquired D by promising to use D, i.e. $-D$ to R. R does not need to know the name of the delivery agent, only that it exists, so the body D does not contain that non-local information. It will be up to L itself to introduce itself to R. The receiver accepts the promise of a message, signalling its willingness to accept such a message $-M$. It doesn't have to accept the condition, since this it has no influence over that decision.

Now there is a link agent L which promises delivery service $+D$ to S, which S accepts $-D$. The link agent further promises delivery to R, conditionally on getting a message from S, i.e. $+D|M$. Its conditional role is the reverse of that for S, and the diagram shows the symmetry between S and L as they exchange roles. S also promises to deliver a package to L, i.e. $+M$, and L accepts with $-M$, unconditionally.

There is now a deadlock of precedence relationships promised, whose order can only be broken by an explicit act of symmetry breaking (i.e. a boundary condition at S).

$$S \xrightarrow{+M} \blacksquare\, L, \tag{16.49}$$

i.e. an unequivocal change of state between the sender S and the delivery agent L. A symmetrical imposition by the receiver does not return the message to sender:

$$R \xrightarrow{+\overline{D}} \blacksquare\, L, \tag{16.50}$$

because the extended and non-local process has direction built into it. It is unidirectional. It amounts to a signal or tick of its clock 'I am still waiting for a delivery', which has no causal impact on S, the origin of messages (unless the process is of sufficient scale to incorporate processes for customer satisfaction!). This underlines the fact that (in spite of the complementarity of give and take $\pm$) it takes more to create reversible system, because every point location or agent is a priori independent in its choices. This is the result of putting locality first as an organizing principle.

Considering then the promises in figure 16.2, the main focus is the promise from sender to receiver:

$$S \xrightarrow{+M|D} R \qquad (16.51)$$

This promise, by itself is an empty one. It need to be completed according the rule for assisted promising (section 6.2). We leave that as an exercise for the reader (shown in the figure). The complementary interpretation, which shows the exchange of (+) and (-) promises is also shown in the figure, by the fully symmetrical transformation:

$$\overline{-M|D} \;\rightarrow\; -\overline{M}|\emptyset \qquad (16.52)$$
$$+\overline{M} \;\rightarrow\; +\overline{M}|\overline{D} \qquad (16.53)$$
$$\overline{-D|M} \;\rightarrow\; -\overline{D}|\emptyset \qquad (16.54)$$
$$+\overline{D} \;\rightarrow\; +\overline{D}|\overline{M}. \qquad (16.55)$$

This examination of the role of intermediate agents in propagation tell tells us two things. The implication is that the intermediate or 'link' agent L is irrelevant semantically and may be absorbed by superagent scaling. It also tells us that promising a completely ordered potential in a lattice becomes N^2 expensive as the number of intermediate agents N grows. Even at a macroscopic level, this is expensive and taxes the resources of agents. On an elementary process, it seems highly unlikely that one would find such a structure. Evidence in the chemistry of long chain molecules and metals shows at indistinguishability plays a saving role to prevent conditionality from diverging.

16.2.6 Co-dependence (entanglement)

When agents make promises that render them co-dependent, by virtue of mutually binding conditional promises, made equally in advanced and retarded directions, it no longer makes sense to distinguish the roles of sender and receiver, as both agents are locked into both roles. In this case, we cannot speak of partial order, or even causal order on a microscopic scale. Entanglement is an explicit case of a mathematical *preorder*. The behaviour of co-dependent agents depends on the relative timescales of the processes that keep promises. Here I'll largely follow the more extensive discussion in [BBKK18].

When agents collaborate, or act cooperatively, it is natural to define a 'superagent' to label them as a collective entity[Bur15]. When such a superagent, composed like $S = A_1 \oplus A_2$, makes promises that cannot be attributed to or kept by either of its components A_1 or A_2 alone, then we say the subagents are entangled, and we say that the superagent is irreducible[Bur15]. This happens when promises are mutually

16.2. ADJACENCY BETWEEN AGENTS

conditional. Consider a two agent system, with agents forming a diatomic molecule, which we can label left and right.

Definition 95 (Entangled with respect to promised property b). *Two agents A_L and A_R are said to be entangled or irreducible if the superagent $A_L \oplus A_R$ enveloping both of them makes co-dependent promises about body subject b that neither of the two agents can make alone.*

This definition is compatible with the definition of entanglement in information theory[DCP17].

Lemma 13 (Entanglement is conditional). *This can only happen if each agent makes promises conditionally on promises made by the other.* □

For any promise bodies b_L, b_R, the necessary and sufficient solution to this condition is given by

$$A_L \xrightarrow{+b_L | b_R} A_R \qquad (16.56)$$
$$A_R \xrightarrow{-b_L} A_L \qquad (16.57)$$
$$A_R \xrightarrow{+b_R | b_L} A_L \qquad (16.58)$$
$$A_L \xrightarrow{-b_R} A_R. \qquad (16.59)$$

The proof is trivial: both sides promise b_i ($i = L, R$) with a dependence on the promise $b_{\bar{i}}$ from the other, else they would promise independently which would contradict the definition. If the agents do not promise the explicit dependence on the other in (16.57) and (16.59), then (16.56) and (16.58) are not complete promises, by the conditional promise law 6.2 of [BB19], that no dependent promise can be given without accepting the dependent promise of the other, thus A_L must accept b_R and vice versa.

In spite of a simple proof, the dynamical behaviours of this co-dependent configuration is not completely defined. It depends on the relative timescales for communication and sampling at each end. In physics one would normally assume agent symmetry by default, but that violates the principle of autonomy, and amounts to assuming reversibility[79]. When b_L or b_R changes, these promises may be thought of as cyclicly generating an evolving sequence of preconditions, which unfolds as a chain of transaction events, until an equilibrium is possibly reached.

16.2.7 BOUNDARIES AND HOLES

The interruption of continuity in a structure of positional agents x_i is what we mean by a boundary. If a superagent structure can no longer promise that there is anything next to it, we have reached its edge. Boundaries explicitly break long range symmetries and seed the formation of structure by anchoring symmetry generators to some fixed points. The simplest notion of a boundary is the absence of a promise of adjacency. For autonomous agents, this can have two possible directions. Suppose one names agents in some sequence; at position n,

$$x_n \xrightarrow{\emptyset} x_{n+1} \tag{16.60}$$

An agent might be ready to accept messages from a neighbour, but there is no neighbour to quench it,

$$x_{n+1} \xrightarrow{+\text{accept presence}} \emptyset \tag{16.61}$$

$$\tag{16.62}$$

or no promise given to keep such a promise.

$$x_{n+1} \xrightarrow{+\text{accept presence}} x_n \tag{16.63}$$

$$x_n \xrightarrow{-\emptyset} x_{n+1} \tag{16.64}$$

We can summarize different cases:

Definition 96 (Continuity boundary). *If an agent x_i does not promise $+\text{adj}_\mu$ to any other agent may be said to be part of a μ-transmission boundary.*

Definition 97 (Observation boundary/event horizon). *If an agent x_i does not promise $-\text{adj}_\mu$ to any other agent may be said to be part of a μ-observation boundary, or event horizon.*

Boundaries can thus be semi-permeable membranes. Such structures are quite common in biology. Boundaries can be localized or extended (see fig. 16.3). Their perceived extent depends on observer semantics, or coordinatization. The absence of an adjacency along a direction labelled μ between agents A and B may be called a μ-boundary, even though there is still a path from A to B via C, in a direction ν.

Boundaries are usually considered to be the discontinuation of a certain degree of freedom or direction. The default state in a network of autonomous agents is boundary.

Definition 98 (Edge boundary). *If an agent x_i does promises $\pm\text{adj}_\mu$ to an agent that does not exist, we may say that x_i belongs to the μ edge of the space.*

Definition 99 (Material boundary). *The edge of a vector region consisting of those agents that uniformly promises material property X.*

16.2. ADJACENCY BETWEEN AGENTS 243

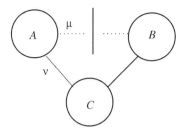

Figure 16.3: A boundary or a partition in a graph may be the absence of an adjacency.

Example 153 (Semantics of boundaries). *Boundaries both semantic and dynamic occur with a rich variety of interpretations.*

- *The absence of an agent adjacency (an edge of space or a crystal vacancy).*

- *An agent that selectively refuses a promise to one or more agents (a semantic barrier, e.g. a firewall, passport control, etc).*

For example:

- *The edge of space itself.*

- *The edge of a property e.g. blueness, occupancy, tenancy, table.*

- *The edge of an organization e.g. firewall 'DMZ'.*

16.2.8 CONTAINMENT WITHIN REGIONS

How shall we represent the idea of one object being inside another in a world of autonomous agents? Agents are atomic, and one atom cannot be inside another. The clue to this lies is viewing containment as a bulk material property. We start by defining membership in regions or cliques.

A compound agent, denoted $\{A_i\}$, with role R is the set of agents that mutually promise to belong to set R. Membership in a group, role of property follows the matroid construction (see section 4.5.8). Containment and overlap may now be defined with reference to fig. 16.4

Definition 100 (Containment promise). *Compound agent $\{A_1\}$ is R-inside (or R-contained by) compound agent $\{A_2\}$ iff*

$$\{A_1\} \xrightarrow{-R} \{A_2\} \tag{16.65}$$
$$\{A_2\} \xrightarrow{+R} \{A_1\} \tag{16.66}$$

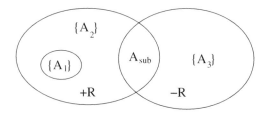

Figure 16.4: Containment and overlap between superagents.

The promise of $+R$ represents membership in region R, which defines a compound or superagent, that is a coarser grain of space than the component agents.

Definition 101 (Spatial overlap promise). *Compound agent $\{A_1\}$ R-overlaps with compound agent $\{A_2\}$ iff*

$$\{A_{\text{sub}}\} \subseteq \{A_3\} \xrightarrow{-R} \{A_2\} \tag{16.67}$$

$$\{A_2\} \xrightarrow{+R} \{A_{\text{sub}}\} \tag{16.68}$$

When we say one region is inside another, it is sometimes convenient to describe the boundary of the region rather than the region's name. There is a straightforward relationship between these, given that a region is simply a material boundary in which there exist adjacent agents, some of which promise $-R$ (inside the region's boundary) and some of which do not (outside the region). The same criterion of membership applies. Using the container as a label for the type membership is the complement of using the type itself, since defining an edge requires us to specify edge with respect to which property.

Example 154 (Computer network hierarchy). *Compound agents inside one another or adjacent to one another can all be represented as regular tuple coordinates, by defining matroids as indicated here. Data network addresses, like IP addresses, VLAN numbers, etc[BBCD14].*

16.3 Symmetry of Short and Long Range Order

Adjacency promise bindings are exactly analogous to chemical bonds, with the addition of semantic types. A graph of autonomous agents with only without adjacency bindings has a state of maximal symmetry; we may call it 'gaseous' or 'disordered', by analogy. If agents promise adjacency promises in a uniform and homogeneous way, one may speak of long-range order as in a 'solid'. It is natural to describe such a space with a lattice coordinate system (e.g. Cartesian), like that of a crystal. For something we take for granted, this is quite non-trivial.

If we are considering spacetimes of technological origin, e.g. computing infrastructure, then we readily see a mixture of these two states in everyday life. Mobile phones, pads and computers migrate without fixed adjacency on a background of more permanent fixtures: servers, disks, network switches, and other 'boxes'. Thus, we should be prepared to view spacetimes of mixed phase within a single picture.

The self-organization of autonomous agents is quite analogous to a phase transition by local interaction in matter. A spacetime with a graph structure thus exists in phases analogous to gaseous (disordered) with short-lived adjacencies, or crystalline with long-lived adjacencies (ordered). The disordered state is said to be a symmetrical state, and a crystalline order is a broken or reduced symmetry, which is usually related to breakdown to arbitrary symmetry operations, by selection of a subset that appears to bring long range order. A lattice exhibits long range order, while a fluid, replete with vortices and flows, might exhibit only short range (nearest neighbour) order, for instance.

16.4 MOTION IN AGENT SPACE

What does motion mean in an agent-based spacetime. in an agent-based spacetime. Three distinct models of motion make sense from the perspective of promise theory. In the first case, there is only a single kind of agent in a gaseous state. In the second, there is a two-phase model with a solid spatial lattice and material properties bonded loosely to them at certain locations. Finally, in the third model, there is only a single kind of agent, but the physical properties promised as matter can bind to a specific agent and be transferred from one to another.

Example 155 (Semantic motion in technology). *Technologically, we have need for all three models of motion. The first relates to ad hoc mobile agents, the second to base-station attachment of mobile devices, and the third to fixed (virtual) infrastructure. It is also fascinating to speculate as to the meaning of these processes in nature.*

In a space of autonomous agents, even the familiar concepts of uniform motion in a straight line are non-trivial without the concept of virtual properties. Through technology we see that the most natural way to reconstruct familiar Newtonian ideas about motion is through virtual processes, i.e. layers of abstraction that build on top of primitive process agents.

16.4.1 MOTION OF THE FIRST KIND

The first case deals with a homogeneous collection of agents that can move by swapping places, within an ordered graph of adjacencies (see fig. 16.5). This becomes increasingly complex in a multi-dimensional lattice, so we'll restrict this to one dimension only to understand its properties.

In order to be able to form a new adjacency, an agent must know about the existence of the agent to which is wants to bind, and vice versa. This can be assumed for nearest neighbours only. However, it turns out that exchanging places[80] also requires knowledge of next-nearest neighbours. This would seem to involve multiple messages back and forth to discover one another. This in turn seems to create a bootstrap problem for spacetime. How can spacetime form structure without such structure existing to begin with? However, as long as agents promise to relay information about their adjacencies to their neighbours, this can be handled in a purely local manner[81].

Consider a simple one dimensional model with adjacencies to the left and right of each agent (see fig. 16.5). We denote a trial agent that has a non-zero velocity from left to right by A_i ($A_i = B$ in the figure). A_{i+1} is to the right of A_i, and A_{i-1} is to the left of A_i. The agent must be able to distinguish left from right.

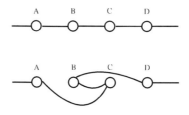

Figure 16.5: Motion of the first kind: intrinsic motion in an untyped spacetime.

In fig. 16.5 we see that, in order for an agent to move one place the right, four agents' promises need to be coordinated. This is a non-local phenomenon, since $A_i = B$ and $A_{i+2} = D$ are initially non-adjacent.

We may denote the bodies for adjacency promises by $\pm\text{adj}_{L,R}$. In addition we denote the property of momentum to the left or right by $+p_{L,R}$. We need to introduce a value function on the agents' promises to allow one combination of promises to be preferred over another $\pi_1 > \pi_2$. Two promises will be considered incompatible when we write $\pi_1 \# \pi_2$. If two promises are incompatible and the first is preferred, i.e. $\pi_1 > \#\pi_2$ then we may assume a promise transition in which π is withdrawn in favour of π_1.

Motion from left to right occurs with the following promises, for all i:

16.4. MOTION IN AGENT SPACE

1. $A_i \xrightarrow{+p_R} *$, i.e. we assume an agent has the property of right momentum, which is promised to all observers.

2. $A_i \xrightarrow{\pm A_{i\pm 1}} A_{i\pm 1}$, i.e. all agents mutually promise their neighbours to relay information about their neighbours to them, both $\pm$.

3. $p_R > \# - \text{adj}_R$ the acceptance of a right-adjacency (from the left) is incompatible with the attribute of right momentum. i.e. an agent with right momentum would immediately withdraw its promise of right adjacency. This allows B to drop its adjacency to A in the figure.

4. If the adjacency with right neighbour A_i goes away, try the next neighbour A_{i+1}:

$$A_{i-1} \xrightarrow{+\text{adj}_R | \neg -\text{adj}_R} A_{i+1}$$

This allows A to connect with C in the diagram. Note this requires a memory of the next neighbour's identity, which means that spacetime has to store more than information about attributes.

5. To join B with D and reverse links BC. An agent A_i with right momentum $+p_R$ might prefer to bind to A_{i+2} and relabel its offer of right adjacency to A_{i+1} as acceptance of a right adjacent $-\text{adj}_R$ from A_{i+1}.

6. To drop CD, an agent might prefer to use a combined promise of momentum and adjacency from B, than pure adjacency from C, i.e. $p_R, +\text{adj}_R > \# + \text{adj}_R$.

7. To flip the use-right-adjacency of CB to a give right adjacency offer CB, this might be preferred if the $-\text{adj}_R$ is no longer received from D.

8. Finally, all the left adjacencies also need to be connected along with the right adjacencies in a similar manner.

This is a lot of work to move an agent one position in a lattice. It seems unnecessarily Byzantine, in violation of Occam's razor. In physics, one assumes that motion is an intrinsic behaviour of bodies, but here we see that it is a cooperative non-local behaviour. This might be unavoidable in any description. This requires no non-locally transmitted knowledge (i.e. the bootstrapping is self-consistent). However, it is far from satisfactory. Perhaps one might also view such multiple connectivity as a form of tunnelling akin to *entanglement*, in the sense of quantum mechanics[82].

16.4.2 MOTION OF THE SECOND KIND

In the first approach to motion, there is only a single kind of agent, which suggests great simplicity, but this leads to a highly complex set of behaviours to explain motion. A second kind of motion may be explained by separating agents into two classes, the first of which we may refer to as spatial skeleton agents S_i, which account for the ordered structure of spacetime, and a second kind of agents which promise non-ordered material properties M_j. Adjacency $\{M\} \xrightarrow{\pm \text{adj}} \{S\}$ accounts for the location of 'matter' within 'space', and matter becomes effectively a container for material promises.

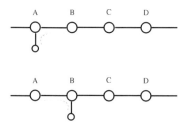

Figure 16.6: Motion of the second kind: extrinsic motion on a typed spacetime.

Motion within such a model then consists of re-binding material agents at new spacetime locations (see fig. 16.6, with $S_1 = A, S_2 = B$).

$$\text{withdraw} \quad M_1 \xrightarrow{\pm \text{adj}} S_1 \qquad (16.69)$$

$$\text{make} \quad M_1 \xrightarrow{\pm \text{adj}} S_2 \qquad (16.70)$$

This is the way mobile phones attach to different cell base-stations, by re-basing or re-homing of a satellite agent around a fixed cell location. It is also the model of location used by bigraphs. Locations are fixed seeds around which material agents accrete or migrate. Since the property of velocity belongs to the material agent, it must be an autonomous promise to bind to a new site. Just as before, agents promising a velocity need to know of the next location binding point. That information has to be relayed between the spatial agents, just as in the previous section. With two classes of agent, new questions arise: how many M_j can bind to the same S_i?

In this model, motion is still not a behaviour that is purely intrinsic to an agent; it is non-local, as the existence of non-adjacent points is not knowable without information by the skeletal agents being exchanged between A and B. However, in this case only two skeletal agents and one mobile agent need to be involved to transport the mobile agent.

16.4. MOTION IN AGENT SPACE

This is much simpler than motion of the first kind. This model effectively recreates the old idea of absolute spacetime, or 'the æther'.

16.4.3 MOTION OF THE THIRD KIND

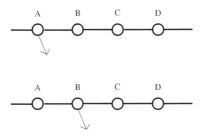

Figure 16.7: Motion of the third kind: extrinsic motion of promises on a non-typed spacetime background.

To remove the notion of an explicit æther, one could avoid distinguishing a separate class of agent, and keep all agents the same. In technology, this is what is known as a peer to peer network. Movement of promises between agents is the third and perhaps simplest possibility. It is essentially the same as motion of the second kind, without a second type of agent but with two classes of promise.

Motion now consists of transferring promises from one location to the next. This brings new assumptions of global homogeneity: now every agent in space must be capable of keeping every kind of promise. There can be no specialization.

The promises to do this are straightforward, and the continuing need for global cooperation with local information in this picture looks very like Schwinger's construction of quantum fields[Sch51, Sch53], and Feynman's equivalent graphical approach[Fey49, Dys49].

Chapter 17

CFEngine: A Promise Keeping Engine

An important chapter in the development of Promise Theory was the development of CFEngine[Bur95, BR97, Bur98]. Many of the principles and realizations that have been formalized originated in the experiences of building a distributed computer system. CFEngine is a software tool that automates the building and maintenance of networks of computers at the software level. It assures the configurations of computers: installing and customizing their software for specific purposes, hardening their settings against security threats, ensuring that only the right programs are running, and a variety of other administrative tasks associated with computer management.

The story of promises, as we describe it in this book, has been intimately related to CFEngine. In fact the importance of promises was first realized in 2004 while trying to find a model that could describe CFEngine's mode of operation[Bur13]. CFEngine uses a model of completely decentralized and distributed management, but most management frameworks that existed at the time of its inception were fully centralized, authoritative remote-control consoles.

With CFEngine running, every computer becomes responsible for its own 'state of health', and cooperates loosely with neighbouring computers if and when necessary. The term 'computer immune system' was used, at one time, as a metaphor for this form of self-sufficiency. CFEngine solves a problem that centralized systems cannot easily solve, by providing a massively scalable parallel system of configuration, and thus it offers a convincing case study for the concepts in this book.

In practical terms, CFEngine combines the notions of promises with the ability to represent patterns (grammars) of an alphabet of promises. In this way, it models

17.1 Policy with autonomy

configurations that can be promised and kept autonomically.

17.1 Policy with autonomy

CFEngine implements computer policy. by a policy we mean the ability to assert constraints on the behavior of objects and agents in a computer system. The most general kind of system one can construct is a collection of objects, each with its own attributes, and each with its own policy. A policy can also be quite general: e.g.

- Policy about behavior,
- Policy about configuration, or
- Policy about interactions with others.

With a promise model, and a network of *autonomous* systems, each agent is only concerned with assertions about its own policy; no external agent can tell it what to do, without its consent. This is the crucial difference between autonomy and authoritative centralized management. Autonomy means that no agent can force any other agent to accept or transmit information, alter its state, or otherwise change its behavior.

17.2 History

Before 2004 CFEngine, like most computer programs, had been written based mainly on intuition of the problem it was trying to solve – namely computer configuration in the highly diverse environment of a university, where many users had special needs to be taken care of. However, after many years of ad hoc experimentation, it became necessary to formalize a model for distributed resource configuration to bring order to this irregular intuition. Curiously, no such suitable model existed in computer science, as distributed resource management is not something that has been modelled traditionally. Computer science tends to treat problems as task graphs, or computer programs to be executed, not as landscapes of multifaceted properties to be described. Promises were the answer to unifying many of the concepts in CFEngine.

The Promise Theory described in this book was originally invented to discuss the issues surrounding this autonomous operation, and voluntary cooperation. Unlike other modeling techniques, like Petri nets or UML, Promise Theory is not about the stepwise development of a device. It is not about protocol modeling, rather it is about *equilibria*, i.e. how to describe steady state behavior that has some underlying dynamics.

Promise Theory begins with the idea of completely autonomous agents that interact through the promises they make to one another. It is therefore particularly well-suited to modeling CFEngine.

CFEngine maintains the principle of *host autonomy*. Many existing systems and descriptions of configuration management assume a centralized authority to change all the parts of the system. This has a number of disadvantages. Our aim here is to begin in the opposite manner, assuming *no centralization* of authority, and then seeing how collaborative structures emerge by free choice.

Promise Theory applies to CFEngine at a number of levels:

- in the way the software is designed.
- in the way parts of the software cooperate.
- in the model used by the software to describe the system.

To apply a promise method to any of these areas, we observe the basic principles and assumptions of the theory. Applying Promise Theory to a representation of Promise Theory is a strangely self-referential thing to do, but it is the only way to be consistent about the principle. It also shows that the idea of promises can be applied recursively at multiple levels in a description of function.

We must begin by identifying the atomic 'agents' in the software along with the promises they make, as well as the atomic agents that are described within the software – i.e. in its model of a computer system through an internal representation of promise statements. Put simply, this involves the following points:

1. identify as many of the resources and their properties, i.e. things belonging to the domain of interest that can be promised independently. This points to a set of smallest independent entities that can make promises, and already gives some clues about necessary and sufficient distinctions.

   This principle can be applied to the computer systems that CFEngine's promises to configure, as well as to the elements in the language that models them, and the individual programs that form the software suite.

2. separate the different types of promises they can make.

   This allows us to define types of promise that can be grouped together for readability in the language representation.

3. identify the details of the promise body for each type. To do this, we re-apply the entire method recursively; to identify the independent issues, or body-parts within the body of a single promise. We can do this because each language representation

of a promise is only a promise statement, and this statement must promise to be a clear and adequate representation of the promise it represents! self consistency is a key here.

4. re-group the independent agents/resources according to those that play simular roles and give the group or pattern a name for easy reference, i.e. Identify which promises belong together to make the best use of patterns that reduce the amount of information used in the description.

5. use patterns to make generic promises that optimize the length of a promise expression.

This was the method used to generate the syntax for CFEngine's promise language, and it forms a method that can be re-used at all levels to bring about an atomization of a problem into necessary and sufficient sized parts.

17.3 A LANGUAGE OF CONFIGURATION PROMISES

Suppose it were decided that a file should have a particular owner. In a declarative specification of this property, one might say something like:

```
files:

mycomputer::

  "/directory/file"

    owner => ''fred'';
```

This assertion says that, in the context of the computer 'mycomputer', the named file in the named location should have the attribute owner with value 'fred'. this is a specification of desired state, and it is in fact what we would now call a *promise* about the desired state of the computer.

Here is a specific example from CFEngine itself. The following is a promise to create a file if it does not already exist, and that the permissions of that file will have certain specified values.

```
files:

  "/home/mark/tmp/test_plain"
    perms => mog("644", "root", "wheel"),
    create => "true";
```

the language is further generalizable. Suppose CFEngine branched out into nano-technology and could build molecules, we might write a configuration to put together something like a water molecule.

```
molecules:

  "water"

    atoms => { hydrogen, hydrogen, oxygen };
```

The language can be understood as a way of abstracting away the details of *how* promises might be kept, i.e. what specific actions that need to be taken in a given situation, from a pure description of *what* promise needs to be kept. In this way, CFEngine reduces a context specific set of processes into a simple description of the desired end state. This is a highly compact form of description that places the focus on intended outcome rather than unknown starting point.

CFEngine's promise language has a generic and extensible form, as well as a fixed grammar, thanks to the use of the promise model. although it could easily be adapted to make any kind of promise, CFEngine uses it to describe the configurations of computers. the remainder of this chapter explains how promises are used in CFEngine.

17.4 THE SOFTWARE AGENTS

CFEngine consists of a number of software agents that contribute to the cooperative system for keeping promises. The role of these agents in keeping promises is explained in the subsequent sections. the main agents are as follows:

- `cf-agent` is the active agent in CFEngine which promises to read a collection of promise proposals (see section 3.3) about computer configuration for an entire network of computers, organized into named bundles. it further promises to pick out those promises that resources (promisers) on the same computer need to keep, and furthermore attempt to keep those promises on behalf of the respective promisers described in the proposals.

- `cf-serverd` is the server 'daemon' or service component[83], which promises to read bundles of promise proposals to grant access to file resources on a given computer. The server daemon further promises to mediate access to those file resources to authorized computers that need to download them in the course of keeping a 'files' promise.

- `cf-monitord`

is the monitor daemon which promises to record and learn the patterns of resource usage on a computer, and then classify the current observed state in relation to the average learned state to inform about anomalous behaviour.

- `cf-execd`

    is the exec' (executor) daemon, which promises to execute `cf-agent` promise-keeping activities according to a predefined schedule.

- `cf-know` is the knowledge agent, which promises to read bundles of knowledge-related promise statements and compile these into a database, structured as a topic map (a form of electronic index). this includes a model of the global promise landscape for all the computers that share the same set of promise proposals.

17.5 The syntax of promises in CFEngine

Promises were used to redesign the syntax of the description language for computers' desired state. The syntax of CFEngine's representation of promise statements grew from a pragmatic mixture of the pre-existing declarative language used by CFEngine before 2004 and the theory developed and described in this book. Today, the CFEngine description language is a more or less clean implementation of the μ-Promise Theory described in this book.

Every promise has a promiser, and a body, which contains a type that refers to the basic configurable objects of the system (files, processes, storage, software packages, etc.). The type is a convenient header to group together promises of the same type under. Thus a generic promise, in CFEngine, looks like this:

```
promise-type:

    scope::

     promiser -> { promisee(s) },

        # body or promise follows in remaining lines

        property => value,
        property => value,
        ...
        property => value;
```

Some examples help to see the domain-specific attributes and their allowed values in action:

```
files:

  "/tmp/test_plain"                     # promiser
                                        # (implicit promisee)
     perms  => mo("644","root"),        # body
     create => "true";                  # body

processes:

  "snmpd" -> { "person","entity" },     # promiser -> promisees

     signals => { "term","kill" };      # body

packages:

  "apache2"                             # promiser

     package_policy => "add",           # body
     package_method => generic;         # body
```

While the precise details of these promise bodies are unimportant to readers who don't know CFEngine, the principles behind them are easily recognizable. The μ-promise model renders all of these diverse matters into a simple pattern so that everything about a system's desired state can be written in this form.

In the example above, the promise types are `files`, `processes` and `packages`. This type is a short explanation of the main intention of the promise, to be kept by the promiser, and detailed by attributes or property assignments in the body (containing =>) implicitly say what the type of promise is.

defining a variable is also a promise:

```
vars:
  "myvar " string => "string me along";
  "userset" slist => { "user1", "user2", "user3" };
```

indeed, everything represented in the CFEngine language is to be thought of as a promise.

17.6 HOW CFENGINE KEEPS PROMISES

CFEngine keeps two kinds of promises:

- promises that are hard-coded into its making and cannot be altered. These promises define the behaviour of the software itself, its defaults and decisions.

- promises that are read in as proposals from a trusted source and kept if they are deemed in scope.

One of the hard-coded promises that CFEngine keeps is to look for a file of promise proposals in a trusted location. The CFEngine agent reads in this file, if found, which may contain a potentially detailed model of the desired state for one or more computers. The promises described in this description file are only *promise proposals*, in the same way that promises in a contract are only proposals until they have been agreed to. The acts of reading this file constitutes a promise to agree to the proposals (by virtue of the hard-coded promise). The agent cannot be forced to read a file by any outside agent however, so this is an act of voluntary cooperation on the part of the agent.

The set of all promise proposals is usually called a policy, and the state of these promises being kept is called compliance with the policy. All computers can share a single policy, and the orchestration[84] of different roles within it is handled in the same way as a musical score – by labelling certain promises with certain players at certain times. In this case, the parts are configuration promises rather than musical promises and the players are instances of `cf-agent` running on different machines.

To handle the orchestration of parts, each promise is prefixed with a context expression or description of scope that describes which of the agents that should keep the promises. For example, if a promise is intended for all windows machines on Monday's, one might write:

```
files:

windows.Monday::

  "c:\mydirectory"
     create => "true";
```

The line with double colons represents the scope for the promise to apply, or a description of which machines need to verify that this promise has been kept. Each agent evaluates these scope rules individually based on the context in which it finds itself.

Once a `cf-agent` extracts the promises it needs to keep from the total policy, it begins to process them one by one. The agent does not necessarily do this in the order that the promises are described, as the internal algorithms for keeping promises contain some embedded knowledge about how best to be able to keep promises.

Because of the principle that promisers can only make promises about their own behaviour, the promiser must always be the object that would be affected by a change. Thus implementing a promise involves checking the current state of the promiser object (a file, process, package, etc) and comparing it to the promised state by working through the details of the promise body. If the current and desired state match, then no action needs to be taken and the agent simply reports that the promise was kept. If there is a discrepancy, however, the agent may either repair the state of the system or merely warn about the lack of compliance[85].

1. the agent expands any variables and assembles all the parts of the promise body into a single constraint list for verification against the current state of the promiser. the constraint list is a possibly long set of pairs of the form:

   ```
   attribute => value
   ```

2. it checks whether an object matching the promiser can be found.

3. the promised object may or may not be supposed to exist. If the promise says that it should exist, then it must be created. If it should not exist, it must be destroyed. Both creation and deletion operations are 'idempotent', i.e. repeating them does not create or destroy things more than once, so the end result will always be that the object ends up existing or not existing, in accord with the promise.

4. the promiser generally has many complicated properties and attributes whose detailed state is part of the promise description. the agent thus proceeds to examine the current state of the promiser and compare it to the promised state. For each attribute, the question 'is this part of the total promise kept' can be answered by the comparison. Again, there is a choice of whether to simply warn about discrepancies or whether to repair them. This decision too is part of the promise body, e.g.

   ```
   files:

     windows.monday::

       "c:\mydirectory"
           create => "true",
           action => warn_only;
   ```

   The default behaviour is to repair any non-kept parts of the promise body.

17.6. HOW CFENGINE KEEPS PROMISES

the precise details of what a CFEngine agent does in order to repair the state of a system go far beyond the scope of this book, into the realms of computer engineering.

A simple abstraction that can be made to keep promises is the notion of an operation that is 'convergent', i.e. that when repeatedly and unintelligently issued like a mantra will always result in the desired state[Bur04b]. This idea predated the notion of promises, but aligns well with the promise model. We can imagine a promise-keeping operation $\hat{o}_b$ for promise-body b which when applied to a state q_{any} results in the promised state q_{promised}, i.e.

$$\hat{o}_b \, q_{\text{any}} = q_{\text{promised}}. \tag{17.1}$$

of course, the promised state also belongs to the 'any' state category, so we also have:

$$\hat{o}_b \, q_{\text{promised}} = q_{\text{promised}}. \tag{17.2}$$

These two relations constitute what is meant by convergence, and so we can interpret $\hat{o}_b$ as a promise-keeping operator. CFEngine's internal algorithms are based on the strategy of introducing one such operator for each independent promisable attribute of a promiser. Creating these operators clearly requires some domain dependent knowledge in each case, but since the number of different cases is enumerable this is not difficult.

17.6.1 WHO ACTUALLY KEEPS THE PROMISE?

In the foregoing explanation, we chose a pure, Promise Theory viewpoint in which every promiser was considered responsible for keeping its own state promises. This is somewhat fictitious however. CFEngine deals with objects such as files, processes, disks, interfaces, etc, each of which will be represented by an autonomous agent in the sense of Promise Theory. These objects have no capabilities to play an active role in maintaining their own state. what we can show is that this is just a formality, and the modelling is an appropriate fiction, as long as there is a simple trust relationship to the cf-agent program, which has the power to force a local system into compliance.

In CFEngine parlance, promises are actually kept by operators within `cf-agent`. An operator is a software algorithm that makes a change. These algorithms are located within `cf-agent`, not within the promisers (files, processes, etc), as the promisers do not have the capabilities to be self-repaired. However, this is no problem as the agents can always borrow this capability as a service by the powerful external agent. An operator then is easily explained as a bilateral service agreement between promiser and the powerful external agent: the promiser promises to inform the agent of its current value and accept any modifications from the agent. This is a simple indirection that is of mainly formal interest.

17.6.2 How long do promises last?

Promises in CFEngine last as long as they remain in scope. A promise can go out of scope in a number of ways.

- a description of scope, containing a time-dependent expression expires: e.g. consider the promise:

```
reports:

  monday.day22.march::

    "today is monday";
```

this is a promise for CFEngine to report the string "today is monday" only on Mondays when the date happens to be the 22nd of march. If this promise is kept, it would likely be a long time before it would be kept again, by which time the following might have happened.

- a policy gets overwritten by a new set of promise proposals that no longer contain the same promise, at such a time as when the agent voluntarily updates its policy. e.g. the following is a promise to update policy from an external source:

```
files:
  any::
    "$(sys.workdir)/promises.cf"
       copy_from => cp("/source/promises.cf","remote_host");
```

an agent can update its policy when a local owner of the policy decides to change it, or when the local owner chooses to allow the agent to get updated instructions from an outside trusted location.

17.6.3 Coordination

By taking autonomy to its logical conclusion, we make very operation and every atom independent. But some more complex processes require coordination between promises. CFEngine handles this by using a dynamic scope as a method of communication between the agents. As the circumstances and status of machines changes, the scope of certain promises widens or narrows to alter the promises that apply to different agents[86].

17.6. HOW CFENGINE KEEPS PROMISES

The term classes is to some extent a misnomer in this context, born of the fact that there is only a single mechanism for cachine knowledge about the world in CFEngine. Ignoring the name, one may think of classes as a kind of flag mechanism that sets the context for a new promise proposal to be kept.

Here is an example of 'classes'.

```
bundle agent CFEngine_processes
{
vars:

  "components" slist => { "cf-execd",
     "cf-monitord", "cf-serverd", "cf-hub" };

processes:

  "$(components)"

     comment => "make sure server parts of CFEngine are running",
     restart_class => canonify("start_$(component)");

commands:

  "$(sys.workdir)/bin/$(component)"

        comment => "make sure server parts of CFEngine are running",
      ifvarclass => canonify("start_$(components)");

}
```

17.6.4 PROMISE BUNDLES

In CFEngine, promises may be arranged into bundles. These bundles can be parameterized in order to make reusable components. This is a full implementation of the principles for componentization, as described in section 14.5.5.

```
bundle agent mybundle(parameter1, parameter2)
{

}
```

Similarly, promise bodies can be parameterized according to the two strategies for promise variation as described in section 14.5.5. A non-parameterized, or pre-packaged body would be written like this:

```
bundle agent mybundle
{
files:

   "/tmp/file"

      perms => safe_settings;
}

body perms safe_settings
{
mode => "0600";
}
```

This does not expose the choice of settings to the user in the promise itself. The alternative formulation, where the body component is parameterized would look like this:

```
bundle agent mybundle
{
files:

   "/tmp/file"

      perms => safe_settings("0600");
}

body perms safe_settings(parameter)
{
mode => "$(parameter)";
}
```

17.7 CFEngine's design components

One of the practical difficulties of using Promise Theory to describe large scale infrastructure as CFEngine does, lies in coping with patterns or repeated instances of whole bundles of promises. Even the most complex infrastructure has common themes that can and must be reused to compress the total information in the configuration, and this is to our economic advantage. We should not have to reinvent every wheel, every time.

17.7. CFENGINE'S DESIGN COMPONENTS

Consistency and compression of design is achieved by the parsimonious use of themes in an environment. Moreover, there might be multiple players in the system making equivalent or at least interchangeable promises. How do we choose between them?

The design of CFEngine tries to put a pragmatic face on this, and to use Promise Theory principles to unravel a potential mess. The notions of 'techniques' and 'design sketches' have been introduced by different users in order to break down a total system infrastructure into re-usable components. This allows infrastructure to be packaged as a kind of commodity.

We shall not refer to the specific implementations of 'sketches' and 'techniques' here, but rather look at what principles guide the constriction of any set of components of CFEngine promises.

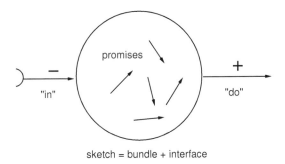

Figure 17.1: CFEngine components are bundles of promises that cooperate to play a particular role in promising infrastructure. They use input from parameters provided by the environment and may other rely conditionally on other promises for keeping their own. The result is an aggregate promise to 'do' something, or represent a role.

Thanks to the basic sufficiency of promise algebra, CFEngine's sketches are little more than parameterized promise bundles, along with a number of guidelines for connecting them into an end-to-end system. By using a tool (the prototype is called `cf-sketch`) to combine sketches into a total design, adhering to the approved guidelines, one attempts to ensure the safe combinatorics of promise patterns. The basic micro-promise formalism makes makes this possible, but not necessarily trivial, as one must always deal with the issue of incomplete information. Naturally, promises can also be used to define these interfaces, so the entire problem is reducible entirely to one of keeping promises at different levels (see section 14).

17.7.1 DESIGN GUIDELINES FOR COMPONENTS

Based on the discussion in section 14, we can summarize some guidelines for the design of components in CFEngine.

Promises relating to correctness:

- Components promise a named with a unique root and a version number.

- Components separate data that will parameterize the components into a separate bundle that can be machine written for tool assistance.

- Components promise to advertise any exclusive promises made. Components should avoid promising to change globally shared resources in an absolute way. Shared resources should only promise minimal, relative state specifications. If that cannot be avoided, then the components should promise to make this information available to other components.

- Components that make conditional promises to use dependencies promise to refer to the specific, acceptable versions of the those dependencies.

Promises relating to usability cost:

- Components promise to have few parameters. Not every possible choice should be parameterized, else there is no benefit to componentization – choose rather more unparameterized components to cover diversity.

- Component names promise to have a relatively flat structure without a complex hierarchy.

As a meta-promise, the number of components should be kept as small as possible to avoid combinatoric complexity when composing them, but this should probably not be done by artificially collecting integrating many promises into large, monolithic, programmable components[87], as this is the opposite strategy to componentization.

17.7.2 AN EXAMPLE FORMAT FOR COMPONENTS

We will abstract an idealized form for the components, without following techniques or design sketches directly. In the example code below, we see an example pattern for a design component or 'sketch'. It has a number of features: a private namespace, and some well-known services or bunde names that represent how it promises to *use* data from outside (the − promise), and *give* service back (the + promise).

```
#
# Design sketch : Check named directories recursively
#

# Make names private to help avoid exclusive promises

body file control
{
namespace => "check_dir_changes";
}

#
# Input of data into clearly understandable variables from tools
#

bundle agent in
{
meta:

  "comment"      string => "Check a list for change (also called a tripwire)";
  "version"      string => "1";
  "depends_on"   slist => { "CFEngine_stdlib" };
  "exclusive_promisers" => { cf_null };

vars:

  "directory_description[/etc]"    string => "System settings";
  "directory_description[/usr]"    string => "System base release";
  "directory_description[/bin]"    string => "System binaries";
  "directory_description[/sbin]"   string => "System operator special binaries";
}

###############################################################################

bundle agent do
{
vars:

  "dir" slist => getindices("check_dir_changes:in.directory_description");

classes:

  "$(dir)_exists" expression => isdir("$(dir)");

files:

   "$(dir)" -> { "goal_infosec" }

        handle => "check_dir_changes",
        comment => "Change monitoring: $(check_dir_changes:in.directory_description[$(file)])",
        changes => default:detect_all_change,
   depth_search => default:recurse("inf"),
   ifvarclass => canonify("$(dir)_exists");

}
```

17.7.3 CONDITIONAL PROMISES AND STACKING OF COMPONENTS

Sketches are containers, like promise bundles, and one is free to design them according to preference. It would be unwieldy for every sketch component to promise all aspects of every problem. Then there would be a lot of overlap between sketches. This is not necessarily a problem as long the promises do not conflict, but it makes sense to try to disentangle resources that can be considered shared into separate agents.

To keep to the promise principles then, the design of sketches ideally (but of course voluntarily) stick to the rule that any promiser that *can* keep its promise independently, should be represented by a different agent. In this case, that means that we would make separate bundles and sketches for separable issues.

Sketches might then still rely on one another (see section 14.5.7). For example, the promise of a database service might be offered by some agent as a direct promise to a user, or as a promise to an intermediary piece of software like a web service. Further, true to the notion that CFEngine is the combination of promises with patterns (grammar), we may try to keep the patterns as regular as possible.

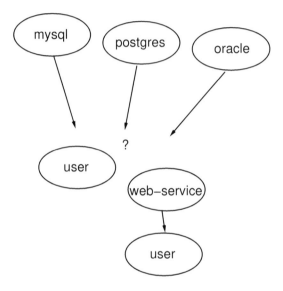

Figure 17.2: If promise bundles play separable roles, then they should form different sketches. Similarly, several sketches might play equivalent roles as far as the promisee/consumer is concerned.

Formally, what we are doing is creating a palette of infrastructure patterns from *conditional promises*. Conditional promises say 'I will promise you b_1 as long as someone keeps their promise of b_2 to me' (see section 6.1). Typical conditions for promising include:

- Needing data or guidance about the local environment.

- Needing component services forming part of a solution, so that multi-agent promises can be kept through cooperation.

17.7. CFENGINE'S DESIGN COMPONENTS

What could have made cooperation potentially complicated in such a scheme is the lack of a simple model for composing patterns. This is where micro-promises' atomic construction comes to the rescue.

17.7.4 COMPOSITION OF COMPONENTS

Readers are referred to section 8.2 for a discussion of promise conflicts, and section 14 on componentization.

There is some ambiguity in the way we define agents or components in a system. This arises from the ability for a collective og agents to represent a role. By cooperating, agents can make promises as if they were a single unit, e.g. an organization. In CFEngine, this means that we might draw the line between single design component and multiple cooperating components as we like. In electronics, one might have the analogy to whether one sells only transistors, resistors and capacitors, or whether one sells integrated amplifiers on a chip.

Design components represent functional issues that span a 'commons' of files and processes resources. The problem of shared system resources, comes into the fore in a major way as systems grow complicated. Some promising agents, like files, networks resources, databases, etc, promise services to multiple clients. This can lead to the possibility of conflict.

For example, a database might promise service as a standalone service, or it could be implicated as a back-end service to a web service (see fig. 17.2).

For example,

- You can't plug two different databases into a web server that only expects one, so a choice has to be made.

- A web service cannot believe a promise by two different intermediate agents to gain exclusive access to a shared resource at the same time, regardless of whether they know about each other.

These basic realities affect how sketches may be combined to provide cooperative promises.

It is tempting to think that any one of the services might do to quench the need for a database, but this is unlikely to be true. Since CFEngine promises are not unique to a single host, but probably span many hosts, searching for what ever happens to be out there would easily lead to *ad hoc* inconsistency, or an unknown desired state.

The key principles that illuminate most of the patterns in Promise Theory are that no agent (in this case no sketch) can promise something on someone else's behalf, and that promises are not guarantees and hence they might not be kept. Thus, when we intend to build cooperative systems, each part of the system has to assume the possibility that dependency promises will not be kept and make this clear to those 'downstream'.

A sketch, or bundle of promises, expecting input parameters from an outside agency thus has to validate things it depends on:

- That all the promises it relies upon from third parties are fully independent, i.e. that they are used to bring about a unique end-state, without conflicts of usage.

- That, even if they don't directly conflict, no two dependency promises relied upon to quench a conditional promise b, make a promise they cannot keep by advertising exclusive access to a third agent's resource, e.g. two webservices that promise exclusive use of the same database. This would lead to contention for the resource that could not be detected directly by the agent promising b. It could however be inferred from accurate promises made by the dependencies.

These are the kinds of cooperation issues that are involved in trying to componentize a complex system. The situation is not so different from trying to design off the shelf components for anything: electronics, furniture, etc.

17.7.5 ACCURATE NAMING OF SKETCH BEHAVIOUR

The naming of sketch dependencies is equivalent to the accurate naming of a conditional promise made by the sketch. It needs to reflect its behaviour, and the behaviour it relies on from promises made by dependency sketches. To create proper expectations, a sketch should thus announce the conditionality of its promised behaviour to its promisees (see section 6.2.1).

We see that the accuracy of the expectations depends very much on the quality and specificity of the promises themselves. Sketches thus have to promise to advertise their services accurately in order to be successful at composition of services in an environment of limited scope.

It was tempting to try to form a taxonomy of sketches based on file names, however file hierarchies are mutually exclusive locations, which are poorly suited to promise behaviour[Bur12]. A hierarchy tries to put every sketch into a unique category that explains it. However, most categories are in the eye of the beholder and are at best ambiguous[88]. A flatter structure was suggested by both by CFEngine's history and by micro-promise principles.

CFEngine's existing approach fro this is to classifying properties, such as in classes promises that classify or tag the properties of environments. Each sketch promises certain meta-data 'tags' that announce and classify its intended purpose. This smoothly allows for multiple sketches to co-exist providing similar or even identical promises.

In a promise view of the world, agents cannot be forbidden from making the same promises, it is up to the consumer of such promises to ensure that it has made a good choice.

CHAPTER NOTES

NOTES

[1] Many accounts of promises are tied to the legal profession and the notion of whether or not promises should have been kept. They intertwine promises with obligations.

[2] The natural sciences, like physics, chemistry and biology, are based on the idea that there are unique and constant facts (often called 'laws' of nature) that can be discovered. These are essentially promises made by the things concerned, as there is nothing that obliges them to be that way. What then remains in those subjects is to explain how the things of the world behave and change, which is called dynamics.

In the human realm, we have to add a new dimension that does not exist in nature: namely *purpose* of things, or the *intention* behind them. This is a purely human perspective. When we make tools, we do so with a purpose in mind. Nature has no purpose either for stone or metal, but human intentions shape the world into patterns that could be the subject of promises.

If we had a theory of behaviour that included promises, it would have to include physics, because physics is just the dynamical part of the theory of behaviour for constant intent (and no purpose). Promise Theory's goal is thus to be a generalization of methods of physics to describe things that happen, applied to the human world of purpose and intent. That is too ambitious for an introduction, but we can try to focus on how Promise Theory deals with the semantic aspects of behaviour that physics cannot.

[3] We can see that a plant is red, or has thorns. A light-bulb might have a wattage rating, or a certain size. A door handle signals whether a door opens inwards or outwards.

[4] The arguments for this are essentially a religious tradition that grew in the past and persist even now.

[5] The notion of a promise can be divorced entirely from the subject of morality and obligation, however. There exists a perfectly consistent and indeed simpler theory that only concerns the management of expectations. An obligation is a sense of foreboding about potential consequences of not acting, and adding such a sense of morality adds compounding layers of subjectivity to an idea that is already bursting with it; moreover, it limits the discussion to promises that refer to future actions. It is possible to promise that something happened in the past, which is difficult to understand from a moral perspective. Rather such promises are about strengthening expectations and trust. The benefit of our somewhat pragmatic approach, we believe, is that it lends itself to a rudimentary mathematical formulation, thus offering a particular kind of clarity usually withheld for engineering disciplines.

The moral discussion of promises is often traced back to the Scottish philosopher David Hume[Qui98, Hum78], who saw promises as something of a mystery. If a promise confers an obligation (to keep it) in the view of society then Hume believed this was paradoxical as one cannot oblige oneself to obey. According to our definitions, Hume's apparent paradox is resolved by a discussion of the *scope* of a promise, i.e. the set of individuals who know about a promise (see next chapter).

A privately held intention, such as Hume imagines, certainly carries no sense of personal foreboding, as there is no independently rationalizable penalty to withdrawing it. However, once an intention is made public (with wider scope than the originator of the promise) the intention becomes a promise proper, and the public knowledge may invoke feelings of peer-admonition should the promiser renege on his or her word. By developing this notion of scope, we avoid bringing morality into the discussion – though, this example suggests one way in which the moral discussion might derive from more elementary considerations.

Another problem with the moral dimension to promises is that one cannot transfer such a notion of promises to inhuman objects, as one frequently does in normal conversation, e.g. 'the weather promises to be rather inclement this morning', or 'the promise of happiness awaits us'.

[6] Why are obligations discussed so much and promises so little? The reasons for this are interesting in their own right. We can only speculate on them here. Probably the concept of obligation is tied to the way in which humans govern, top down. The Law, starting with perhaps the Magna Carta, set a norm for centralized governance by a few powerful individuals who exerted organization or influence over a large number of others. As time has passed, desirable behaviour has become the norm and attention naturally refocuses on individuals' voluntary cooperation rather than top down force and obligation.

[7] In computer science, particularly the field of Multi-Agent Systems the concept of *commitments* has been used for some time [WS03, Woo02]. It has been suggested that promises and commitments are the same. However, we shall show that this is not the case. More seriously, the sense in which the term commitment is used in such discussions is more stylized than purposely considered.

[8] The phrase 'exact science' is of course pure nonsense. As Hume pointed out, there are two kinds of knowledge: provable assertions about simplified systems where all assumptions are laid out (i.e. a fictitious world), and unavoidably approximate or even ad hoc observations and derivations of the real world. In both cases there is is uncertainty.

[9] We are not talking about the moving parts of a clock, or even an Internet search request, but the motivations that play a role in making everything happen. Actions may or may not be necessary to fulfill intentions. Maybe inaction is necessary!

[10] Thus the iterated observation of promise networks maps to Baysian probability networks.

[11] We deliberately use the spelling 'promiser' in this book, rather than the more usual 'promisor', in order to keep in mind the separation between a voluntary agent and one that is bound by a 'promissory obligation' as in other works.

[12] Atiyah [Ati81] suggests that any promise leads to an obligation to keep that promise that is motivated by the threat of tit for tat reprisals. Reciprocation is thus coupled to the idea of promises immediately, which seems to hop over fundamental definitions directly to a discussion of the economics of keeping promises. The obligations are to avoid injury and to reciprocate goodwill. It might be discussed whether incentives are the same as obligations. Atiyah points out however that promising something cannot be necessarily used to create obligation at will. Promises might cause obligations but they can also represent obligations that already exist, i.e. to show commitment to an existing obligation to pay the price of something. e.g. I promise to pay the bearer the sum of 1 pound (in gold). This is only an existing admission of moral obligation. Atiyah maintains, plausibly, that the motivation for promising has changed throughout history. When people make promises, their intentions are culturally bound. Thus a Victorian gentleman's conception of a promise might not fit with that of a present-day child who promises to be home in time for dinner.

Cartwright takes Atiyah's view and asks what might be the point of promises if not to generate the assumed obligation [Car84]. Why do people bother to make promises about things to which they are already obliged? His answer includes the idea that it is a face-saving measure: to mitigate their humility, suggesting that an obligation is interpreted as a kind of attack or levy of force? Alternatively, perhaps the obligation to keep one's promises weighs heavier than the original obligation (I promise you my word as a gentleman not to kill you, even though the law says I am forbidden). Referring to Fried [Fri81], Cartwright points out that the economics of contractual tit-for-tat suggested by Atiyah is tied to promises and not to the obligations they might confer.

The idea that promises are an economic driver of contracts or agreements as bilateral exchanges of promises is continued in the work of Gilbert [Gil93]. Then Carrillo and Dewatripont have argued that promises can best be understood as a market mechanism for reducing the uncertainty in a moral-hazard game [CD08]. This work does not seem to have been pursued. Does a promise increase the likelihood of voluntary cooperation? A number of other works mention the concept of promises in the context of Game Theory also. In these, the concept of a promise is tacitly assumed to be related to the probability of choosing a particular game strategy.

Scanlon [Sca90] meticulously analyses how and to what extent promises give rise to obligations under a variety of combinations of additional assumptions. In his analysis morality plays a important role and it is implicitly assumed that promiser and promisee are human beings capable of moral reflection.

Zhao et al. [ea05] provide a comprehensive modal logic incorporating beliefs, capabilities and promises. Unfortunately it is difficult from that work to extant a clear intuition of the concept of a promise that the authors had in mind. It seems that this difficulty is in part caused by the formalist approach taken. In Framinan and Leisten [FL10] order promising is displayed as a standard technical term in industrial workflow management, while at the same time that use of the term promise is considered lacking a sufficiently clear definition.

[13] All descriptions of deontic logic are fraught with logical difficulties.

[14] For a discussion of these topics, see In Search of Certainty by Mark Burgess [Bur13].

[15] There might conceivably be other models of promises, with different agendas. We shall not attempt to look for these. The model described here has been designed to be used in an engineering or design context.

[16] Promises that are indistinguishable have no purpose because, as we shall show below, they are mathematically idempotent. Thus identical promises with the same name (or without a name) are simply a reiteration of the same promise. By using names, one could artificially distinguish between identical promises for the sake of a larger coordination of process.

[17] The classical theories of relativity by Galileo and Einstein deal with all the same issues as we must deal with here for continuum models. A model of agents is a discrete network so the notions of speed of information are not defined in the same way.

[18] The existence of possible worlds was postulated by Leibniz, Kripke and Everett. For a review, see [Bur13]

[19] Promises bear a passing resemblance to the theory of capabilities in ref. [Sny81]

[20] In earlier editions, we used the opposite convention for $\pm$ for impositions. This later felt inconsistent, and the present labelling is preferred, for reasons that especially have to do with the complementarity discussed below.

[21] Such is it with all physical laws and mathematical expressions of change that there is no implied direction to these causative arrows initially. It is up to the analyst to break the symmetry by specifying a *boundary condition* (or, in this case, initial condition), which manifestly breaks the symmetry by deciding a given point at which we have certain knowledge of the system concerned and in which direction from this milestone the predictions (promises) of behaviour take us from that point.

[22] Notice that the decomposition is not necessarily a hierarchy (or particular spanning tree) – it can be any identification of subset patches that we like. Thus we are not locked into an authoritative taxonomy of roles.

[23] Roles have been defined many times in computer science. Bishop defines them to be groups[Bis02] that tie membership to function. When a subject assumes a role, he or she is given certain skills or rights within the system. This delegation of rights and capabilities is performed by some central monitor, usually a system kernel. A role, according to Canright and Engø[CEM04] on the other hand, is a type of node that defines the topology of the total graph. A role can be source, sink, bridge, a member or a region, or a dangler (a dead end that is both a source and a sink).

[24] In networking, this is sometimes called an 'anycast' proxy.

[25] Looking ahead to the promise life-cycle in figure 7.1, the repeated process of assessment may be identified as the self-loop marking 'kept', 'not kept', etc.

[26] These choices mirror the propagation of data from boundary values by Green functions, e.g. in electrodynamics (see fig. 5.1)[Bur02]

[27] An assessment function may be called linear if it satisfies the relation:

$$\alpha_a(\pi; t_i, t_f; \sum_i c_i M_i) = \sum_i c_i \alpha_a(\pi; t_i, t_f; M_i). \tag{17.3}$$

This form of assessment function can be used as a notion of value for accounting purposes, thus it allows us to associate payoff, or utility to promises in a manner consistent with Game Theory and economics. In particular, we note that a linear combination of sample measurements $\langle M \rangle = \sum_i \delta^i M_i$, where δ^i means raised to the power i, and $\delta < 1$, represents a discounting of value over time, as is common in economic valuations.

Although linear functions are especially important, they are not the only assessments. A form of some importance is the quadratic invariant that represents the Euclidean norm:

$$\mathcal{A} = \sqrt{\alpha_1^2(\pi; t_i, t_f; M_1) + \alpha_2^2(\pi; t_i, t_f; M_2) + \ldots} \tag{17.4}$$

This is used in the theory of errors and uncertainties to compute a fair metric distance for independently assessed uncertainties. It corresponds to the Pythagoras rule.

[28] It is thus a basic process of information transfer[SW49, CT91]. Irrespective of notions of freewill, our model of promises serves a role in making clear the necessary and sufficient conditions for transmission of data. From there, it is possible to address the limitations we must expect to observability. The argument begins along parallel lines to the concept of mutual information in communications theory[SW49]. A measurement is the information that results from that part of what the sender reveals that is collected by the receiver.

[29] This illusion of a flow is maintained by a gradient of intermediate agents that accumulate samples in buffers.

[30] It may be argued that these rules are the basis for understanding emergent swarm behaviour[BF07b].

[31] One might want to call this construction an interface[BP06].

[32] When a conditional promise is made and quenched by an assistant, the 'contact' agent is directly responsible by default. We shall refine this view with alternative semantics later, since this is all a matter of managing the uncertainty of the promise being kept. As soon as we allow rewriting rules of this basic type, it is possible to support multiple solutions for bringing certainty with graded levels of complexity.

[33] Think of knowledge as knowing something as you know a friend. You don't really 'know' them when you first meet, it takes experience to know them well. Knowledge is awareness of information, a promise to pay attention to it, and might include a level of recognition.

[34] In physics, equilibria play a dual role. They are 'macrostates' in which there are no observed changes, and yet they also form the the latent generators of time ordered processes. The so called effective action, related to the 'free energy' in thermodynamics, is the sum of all the different ways that nothing (on average)

NOTES

can happen in the system[Rei65, Abb92, Bur02]. This generates a lot of symmetries. By breaking any one of these symmetries, something will happen. Thus the effective action equilibrium state is the starting point for all the ways a system can change. The variation of an equilibrium subject to time ordered boundary conditions breaks time-reversal symmetry and leads to equations of motion in a single time direction. We can use the same principle in promises.

[35] In the mathematics of difference equations, the solution has two parts: a 'particular integral' which describes the transient response to the initial 'kickstart' condition, and the 'continuous integral' or complementary function that describes the steady state behaviour, once all memory of the transient response has subsided. This is directly analogous to the situation here.

[36] This is completely analogous to the way physical law is expressed in terms of differential or operator equations that are reversible, where in reality only one of the directions is observed. The point is that time itself is not unique – it is defined by an ordered sequence of change. That is why is it possible to construct both 'advanced' and 'retarded' viewpoints for matching phenomena to boundary conditions at different points along a generated sequence.

[37] The notion of a contract is often associated with that of promises[Gil93, Fri81]. Part of the controversy lies in the cooperative dilemma of who promise what first. In our interpretation there is no need to worry about time since all proposals are activated as a single event by mutual signing.

[38] Note that a collection of promises is not necessarily consistent or sustainable. An agreement has to be worked out so that it is of mutual economic benefit to both parties, from each of their perspectives. This, however, has to do with the economic sustainability of the promises. An agent can clearly make a promise that it is in fact incapable of keeping.

[39] Many legal agreements, terms and conditions, may be considered sets of promises that define hypothetical boundary conditions for an interaction between parties or agents. The law is one such example. One expects the promises described in the terms and conditions of the contract will rarely need to be enforced, because one hopes that it will never reach such a point, however the terms define a standard response to such infractions (threats). Legal contracts define what happens mainly at the edge states of the behaviour one hopes to ensue.

[40] However, one could imagine a dispute in which the insurer tried to renege on the content of a contract by claiming it was written by a rogue employee on their company paper, or that it was in conflict with some other policy.

[41] The latter explains why non-functional aspects of product design often contribute to their success in markets.

[42] The Wikipedia defines: "Trust is the belief in the good character of one party, they are believed to seek to fulfill policies, ethical codes, law and their previous promises." Using promises, we can address the some of the deficiencies of these discussions by introducing a meta-model for understanding trust. Our model can be used to explain and describe common trust models like "trusted third parties" and the "web of trust".

[43] There is an extensive literature on trust in computer science[LaP94, McI89, Win96, PJ04, JKD05, HJS04]. Much of it is concerned with generating protocols for the purpose of determining the validity of public keys and other identity tokens, or criticizing these mechanistic views in a wider security perspective. Here we are mainly concerned with general ideas about trust and reputation. We find the recent work of Klüwer and Waaler to be of interest from the viewpoint of logic[KW05, KW06]. These authors present a natural reasoning system about trust which includes the notion of *ordering* by levels of trustworthiness.

The work that seems closest to ours may be found in ref. [BBK94] and ref. [JP05]. Here the authors distinguish between trust and reputation and provide an epidemic-like procedure for valuating the trust based on some inference rules and numerical measures that are essentially reliabilities. The calculation is hence mainly appropriate for a frequentist interpretation of probability. The authors in ref. [BBK94] are unable to distinguish trust about different issues, or relate these in their model. In ref. [JP05], an attempt is made at motivating trust types but the underlying properties of these types is not completely clear.

[44] It is a matter of belief whether one assigns this trust to a promise made by an agent called God.

[45] Business modelling typically assumes that organizations achieve logistical cooperation in a push-based clockwork fashion, as if the organization were a single programmed unit.

[46] This is analogous to the notion that continuity requires a total function of the domain that feeds into it.

[47] Readers are warned not to fall into the trap of thinking of promises as physical network connections simply because of the familiarity of the latter.

[48] This is often called a Quality of Service guarantee.

[49] Even though these are only logical connections, the limitations stand.

[50] This is a more realistic architecture for IT systems.

[51] This can approximated by using a DNS load balancing mechanism in some cases, though the technologies for implementing this are currently limited.

[52] We know that all discrete combinatoric patterns are classified by grammars of the Chomsky hierarchy[LP97, Bur04a], which may be formed from the alphabet of such operators. This is consistent with the concept of lowered entropy and differentiation.

[53] Irreducibility is a property of matrix representations. Here is applies to the promise graph. This is a property of multiple agents and is characterized by the eigenstructure of the promise graph which defines natural regions in the graph[CEM04].

[54] These elementary considerations form the basis for a discussion of cooperation and organizational patterns; it also provides the groundwork for a later definition of institutions and their boundaries[Ost90].

[55] These ionic properties may reach outside the superagent boundary too in some circumstances, allowing oxidation, etc.

[56] In fact, in the case of a radio, one could argue that it is the outer casing which makes the promise of being a radio, and that the other components are tenants of the outer casing agent. I'll return to the issue of tenancy in the latter part of these notes.

[57] A full understanding of this phenomenon requires a discussion of symmetry-breaking, however we can discuss a simplified view. We suggest that this emergence of privilege has a simple explanation in a process of structural 'crystallization' which is seeded by a self-appointed promiser of a collation service.

[58] This is a familiar problem in security where it is used for key distribution[Bis02].

[59] It's interesting that this kind of 'public education' service was common after the second world war, but fell into disuse in the 1980s when many of the objectives became normalized. Today, do we assume normalized behaviour is established without such education? The risk of norms decaying into instability is always there. Maintenance of promises in necessary in all systems[Bur03].

[60] Inviting someone to invite them is almost like an imposition, but through the channels of diplomatic maneuvering.

[61] British readers may take offence at the spelling of offense, which is a constant source of consternation between regional norms.

[62] This assumes that we are talking about promises of the first kind, which are the fundamental promises.

[63] This contradicts the mandatory language of 'inalienable human rights' in law, such as the American constitution's bill of rights—whose amendments are in fact nothing more than promises by the state to its subjects. There are no fundamental rights. There are only fundamental capabilities. It makes no sense to speak of the right for humans to be able to speak about X any more than anyone could grant the right to have wings, to fly, or to breathe under water.

NOTES

[64] The etymology of authority comes from the Latin auctoritas, meaning 'originator, promoter', as in author.

[65] The latter is effectively a form of voting, but not based on a democratic majority. The mandate is required for every single agent individually for cooperation.

[66] As we say in the networking technology world 'No route to destination'.

[67] This is an over-simplification of the antibody mechanism, but the essence is accurate.

[68] This is the promise paraphrasing of 'the customer is always right'

[69] The concept of branding goes back to marks burnt onto livestock. In marketing, branding looks at many aspects of the psychology of naming, as well as how to attract the attention of onlookers to make certain components more attractive to users than others.

[70] This issue mirrors the acceptance of host and user identity in information systems using cryptographic keys which can later be subjected to verification. The cryptographic methods mean nothing until one has an acceptable level of certainty about identity.

[71] In the theory of evolution, older books tend to refer to fitness, or the survival of the fittest. Their argument was that living systems that promise certain attributes fall short of what is expected to survive in an environment. These life-forms thus exhibit the failure of some components to quench survival promises. For instance, I promise to run very fast if my heart will pump enough blood. The failure of a heart to pump enough blood will lead to the failure of the conditional promise. Of course, in evolution there is no conscious design taking place, merely selection through competitive failure.

[72] We alert the reader to a potential confusion of notation here. The symbol π is normally used for a promise. In the context of Game Theory it is often used also for the payoff or utility matrix. We shall therefore use the symbol $U_{ijk...}$ for the payoff or *utility* matrix of an N player game.

[73] We follow Rasmussen in our description of games to avoid repeating well-known definitions[Ras01].

[74] The subject of semantic spacetime is too large to fit into this introductory book, so we summarize a few of the main points here. See [Bur14, Bur15, Bur16, Bur19c] for further details.

[75] Database normalization rules (first normal form)[Dat99, Bur04c] are promises of regularity of form (internal structure on spacetime elements that are tables) but not just any promise tells us about how to traverse from one place to another.

[76] Note, nothing prohibits there being several channels for this communication, with different characteristics. We think immediately of light messages versus entanglement messaging, for instance.

[77] In physics, we associated Long Range Order with phase transitions and symmetry breaking[Gol92].

[78] One can speculate whether entanglement in Quantum Mechanics is such an out of band channel that enables a short circuiting of local.

[79] It's flawed reasoning to assume the conditions for reversibility by default and then claim that it's physical law.

[80] This is essentially a bubble sort algorithm

[81] A second, but related problem, is how to form a coordinate system in a gaseous phase. If one cannot use the sequential, monotonic nature of integer labels, then coordinate names become ad hoc and lose the extrapolative power of a pattern.

[82] What this exercise in promise theory reveals is the logical need for *memory* of non-local information in order to bootstrap motion. Whether spacetime is artificial or not, this structural information conundrum is unavoidable.

[83] A background program is called a daemon in Unix parlance.

[84] The term orchestration has become common in describing distributed cooperation between computer agents. The term is inspired by the meaning from music, in which the players in an orchestra are able to play together by cooperating voluntarily. This cooperation is possible because each agent has a copy of the

musical score (which may be considered a set of promise proposals for the agent to play certain notes, labelled with the scope of the instruments that should play them, e.g. first violins), and by the promise to use and heed hand-signals from a conductor, who is a coordinating agent. the main role of a conductor is to promise to 'wave in' certain players to remind them about when to begin playing, as well as to provide them with signals about loudness in order to balance the lead. In CFEngine, orchestration is possible because each agent has a copy of the set of promise proposals and knows which promises are intended for it, i.e. which are in scope. A single point of coordination may also be used when it is important for agents to synchronize their behaviour in time.

[85] Compliance can never be guaranteed, because there are always issues beyond the control of an agent. a power failure would incapacitate an agent and prevent its promise of cleaning up a disk of junk files, for example. Thus agents comply only on a best-effort basis. Under normal circumstances, however, special algorithms in CFEngine have been designed to maximize the likelihood that promises will be able to be kept. This is called 'convergent behaviour' and is somewhat beyond the scope of the present book[Bur04b].

[86] CFEngine refers to scope or context expressions as 'classes'. we have tried to avoid the use of this term in this book, as the term is overloaded with connotations from a variety of sources.

[87] A VLSI approach to system configuration has no value in a component framework. In effect, the value of components versus simply designing from atomic primitives are mutually exclusive. Some users will choose to write a complete integrated set of promises for their organization, without breaking it into components, but then they would be unlikely to use any components made by others. Component users are typically users wanting to get ready made solutions. If you can make your own, you don't need them.

[88] For example, what would be the correct classification of the secure shell 'ssh'? Does it belong under security, shells, networking, communications, or some other category? In fact it could promise to be a member of any or all of them, and this is indeed what a promise approach would encourage.

BIBLIOGRAPHY

[Abb92]   L.F. Abbott. *Acta Physica Polonica*, B13:33, 1992.

[AR97]   A. Abdul-Rahman. The pgp trust model. *EDI-Forum: the Journal of Electronic Commerce*, 1997.

[Ati81]   P.S. Atiyah. *Promises, Morals and Law*. Clarendon Press, Oxford, 1981.

[Axe84]   R. Axelrod. *The Evolution of Co-operation*. Penguin Books, 1990 (1984).

[Axe97]   R. Axelrod. *The Complexity of Cooperation: Agent-based Models of Competition and Collaboration*. Princeton Studies in Complexity, Princeton, 1997.

[Bal04]   P. Ball. *Critical Mass, How one thing leads to another*. Random House, 2004.

[BB06]   J.A. Bergstra and M. Burgess. Local and global trust based on the concept of promises. Technical report, arXiv.org/abs/0912.4637 [cs.MA], 2006.

[BB13]   J.A. Bergstra and M. Burgess. A static theory of promises. Technical report, arXiv:0810.3294 v5 [cs.MA], 2013.

[BB14]   J.A. Bergstra and M. Burgess. Promises, impositions, and other directionals. Technical report, arXiv:1401.3381 [cs.MA], 2014.

[BB19]   J.A. Bergstra and M. Burgess. *Promise Theory: Principles and Applications (second edition)*. $\chi tAxis$ Press, 2014,2019.

[BBB07]   J. Bergstra, I. Bethke, and M. Burgess. A process algebra framework for promise theory. Technical report, arXiv:0707.0744 [cs.IO], 2007.

[BBCD14]   P. Borrill, M. Burgess, T. Craw, and M. Dvorkin. A promise theory perspective on data networks. *CoRR*, abs/1405.2627, 2014.

[BBCEM10] J. Bjelland, M. Burgess, G. Canright, and K. Eng-Monsen. Eigenvectors of directed graphs and importance scores: dominance, t-rank, and sink remedies. *Data Mining and Knowledge Discovery*, 20(1):98–151, 2010.

[BBJF] K. Begnum, M. Burgess, T.M. Jonassen, and S. Fagernes. Summary of the stability of service level agreements. In *Proceedings of International Policy Workshop 2005*.

[BBK94] T. Beth, M. Borcherding, and B. Klein. Valuation of trust in open networks. In *Proceedings of the European Symposium on Research in Computer Security (ESORICS), LNCS*, volume 875, pages 3–18. Springer, 1994.

[BBKK18] P. Borrill, M. Burgess, A. Karp, and A. Kasuya. Spacetime-entangled networks (i) relativity and observability of stepwise consensus. *arXiv:1807.08549 [cs.DC]*, 2018.

[BC06] M. Burgess and A. Couch. Autonomic computing approximated by fixed point promises. *Proceedings of the 1st IEEE International Workshop on Modelling Autonomic Communications Environments (MACE); Multicon verlag 2006. ISBN 3-930736-05-5*, pages 197–222, 2006.

[BDT99] E. Bonabeau, M. Dorigo, and G. Theraulaz. *Swarm Intelligence: From Natural to Artificial Systems*. Oxford University Press, Oxford, 1999.

[BFa] M. Burgess and S. Fagernes. Pervasive computing management: A model of network policy with local autonomy. *IEEE Transactions on Software Engineering*, page (submitted).

[BFb] M. Burgess and S. Fagernes. Voluntary economic cooperation in policy based management. *IEEE Transactions on Network and Service Management*, page (submitted).

[BF06] M. Burgess and S. Fagernes. Autonomic pervasive computing: A smart mall scenario using promise theory. *Proceedings of the 1st IEEE International Workshop on Modelling Autonomic Communications Environments (MACE); Multicon verlag 2006. ISBN 3-930736-05-5*, pages 133–160, 2006.

[BF07a] M. Burgess and S. Fagernes. Laws of systemic organization and collective behaviour in ensembles. In *Proceedings of MACE 2007*, volume 6 of *Multicon Lecture Notes*. Multicon Verlag, 2007.

[BF07b]   M. Burgess and S. Fagernes. Norms and swarms. *Lecture Notes on Computer Science*, 4543 (Proceedings of the first International Conference on Autonomous Infrastructure and Security (AIMS)):107–118, 2007.

[BG02]   E. Brousseau and J-M. Glachant, editors. *The Economics of Contracts Theory and Applications*. Cambridge University Press, 2002.

[Bis02]   M. Bishop. *Computer Security: Art and Science*. Addison Wesley, New York, 2002.

[BP06]   J.A. Bergstra and A. Ponse. Interface groups for analystical execution architecture. *(preprint)*, 2006.

[BR97]   M. Burgess and R. Ralston. Distributed resource administration using cfengine. *Software practice and experience*, 27:1083, 1997.

[Bura]   M. Burgess. The promise of system configuration.

[Burb]   M. Burgess. Promise you a rose garden.

[Bur95]   M. Burgess. A site configuration engine. *Computing systems (MIT Press: Cambridge MA)*, 8:309, 1995.

[Bur98]   M. Burgess. Automated system administration with feedback regulation. *Software practice and experience*, 28:1519, 1998.

[Bur02]   M. Burgess. *Classical Covariant Fields*. Cambridge University Press, Cambridge, 2002.

[Bur03]   M. Burgess. On the theory of system administration. *Science of Computer Programming*, 49:1, 2003.

[Bur04a]   M. Burgess. *Analytical Network and System Administration — Managing Human-Computer Systems*. J. Wiley & Sons, Chichester, 2004.

[Bur04b]   M. Burgess. Configurable immunity model of evolving configuration management. *Science of Computer Programming*, 51:197, 2004.

[Bur04c]   Mark Burgess. *Analytical Network and System Administration — Managing Human-Computer Systems*. J. Wiley & Sons, Chichester, 2004.

[Bur05]   Mark Burgess. An approach to understanding policy based on autonomy and voluntary cooperation. In *IFIP/IEEE 16th international workshop on distributed systems operations and management (DSOM), in LNCS 3775*, pages 97–108, 2005.

[Bur08]   M. Burgess. Business alignment through the eye-glass of promises. In *Keynote to BDIM workshop and NOMS2008, Brasil*, 2008.

[Bur09]   Mark Burgess. Knowledge management and promises. *Lecture Notes on Computer Science*, 5637:95–107, 2009.

[Bur12]   M. Burgess. *New Research on Knowledge Management Models and Methods.*, chapter What's wrong with knowledge management? The emergence of ontology. Number ISBN 979-953-307-226-4. InTech, 2012.

[Bur13]   M. Burgess. *In Search of Certainty: the science of our information infrastructure*. χtaxis Press, 2013.

[Bur14]   M. Burgess. Spacetimes with semantics (i). *arXiv:1411.5563*, 2014.

[Bur15]   M. Burgess. Spacetimes with semantics (ii). *arXiv.org:1505.01716*, 2015.

[Bur16]   M. Burgess. Spacetimes with semantics (iii). *arXiv:1608.02193*, 2016.

[Bur19a]   M. Burgess. *A Treatise On Systems Volume II: Intentional Systems With Faults, Errors, And Flaws*. $\chi tAxis$ Press, 2017-2019.

[Bur19b]   M. Burgess. *Smart Spacetime*. χtAxis Press, 2019.

[Bur19c]   M. Burgess. The structure of semantic spacetime and causal sets. *In preparation*, 2019.

[Car84]   J.P.W. Cartwright. An evidentiary theory of promises. *Mind (New Series)*, 93(370):230–248, 1984.

[CD08]   J.D. Carrillo and M. Dewatripont. Promises, promises. Technical Report 172782000000000058, UCLA Department of Economics, Levines's Bibliography, 2008.

[CEM04]   G. Canright and K. Engø-Monsen. A natural definition of clusters and roles in undirected graphs. *Science of Computer Programming*, 53:195, 2004.

[CT91]   T.M. Cover and J.A. Thomas. *Elements of Information Theory*. (J.Wiley & Sons., New York), 1991.

[Dat99]   C.J. Date. *Introduction to Database Systems (7th edition)*. Addison Wesley, Reading, MA, 1999.

[DCP17]   G.M. D'Ariano, G. Chiribella, and P. Perinotti. *Quantum Theory From First Principles*. Cambridge, 2017.

[Dys49]   F.J. Dyson. The radiation theories of tomonaga, schwinger and feynman. *Physical Review*, 75:486, 1949.

[ea05]   X. Zhao et. al. Toward a formal theory of belief, capability, and promise incorporating temporal aspect. *LNAI (CEEMAS 2005)*, 3690:296–305, 2005.

[Fey49]   R.P. Feynamn. Space-time approach to quantum electrodynamics. *Physical Review*, 76:769, 1949.

[FL10]   Jose M. Framinam and Rainer Leisten. Available-to-promise (atp) systems: a classification and framework for analysis. *Int. Journal of Production Research*, 48(11):3079–3103, 2010.

[Fri81]   C. Fried. *Contract as promises*. Harvard University Press, 1981.

[Gao09]   W. Gao. Process management and orchestration. Master's thesis, Oslo University and Oslo University College, 2009.

[Gil93]   M. Gilbert. Is an agreement and exchange of promises? *Journal of Philosophy*, 90(12):627–649, 1993.

[Gol92]   N. Goldenfeld. *Lectures On Phase Transitions And The Renormalization Group*. Addison Wesley, 1992.

[HJS04]   T. Dong Huynh, Nicholas R. Jennings, and Nigel R. Shadbolt. Developing an integrated trust and reputation model for open multi-agent systems. In Rino Falcone, Suzanne Barber, Jordi Sabater, and Munindar Singh, editors, *AAMAS-04 Workshop on Trust in Agent Societies*, 2004.

[Hol98]   J.H. Holland. *Emergence: from chaos to order*. Oxford University Press, 1998.

[HPFS02] R. Housley, W. Polk, W. Ford, and D. Solo. Internet x.509 public key infrastructure: Certificate and certificate revocation list (crl) profile. http://tools.ietf.org/html/rfc3280, 2002.

[Hum78]   D. Hume. *Treatise on Human Nature*. Oxford, 1978.

[IT93]   ITU-T. *Open Systems Interconnection - The Directory: Overview of Concepts, models and service. Recommendation X.500*. International Telecommunications Union, Geneva, 1993.

[JKD05] Audun Jøsang, Claudia Keser, and Theo Dimitrakos. Can we manage trust? In *Proceedings of the Third International Conference on Trust Management (iTrust), Versailes*, 2005.

[Joh01] S. Johnson. *Emergence*. Penguin Press, 2001.

[JP05] Audun Jøsang and Simon Pope. Semantic constraints for trust transitivity. In *APCCM '05: Proceedings of the 2nd Asia-Pacific conference on Conceptual modelling*, pages 59–68, Darlinghurst, Australia, Australia, 2005. Australian Computer Society, Inc.

[KW05] J. Klüwer and A. Waaler. Trustworthiness by default, 2005.

[KW06] J. Klüwer and A. Waaler. Relative trustworthiness. In *Formal Aspects in Security and Trust: Third International Workshop, FAST 2005, Newcastle upon Tyne, UK, July 18-19, 2005, Revised Selected Papers, Springer Lecture Notes in Computer Science 3866*, pages 158–170, 2006.

[LaP94] L. LaPadula. A rule-set approach to formal modelling of a trusted computer system. *Computing systems (University of California Press: Berkeley, CA)*, 7:113, 1994.

[LP97] H. Lewis and C. Papadimitriou. *Elements of the Theory of Computation, Second edition*. Prentice Hall, New York, 1997.

[McI89] M.D. McIlroy. Virology 101. *Computing systems (University of California Press: Berkeley, CA)*, 2:173, 1989.

[Mye91] R.B. Myerson. *Game theory: Analysis of Conflict*. (Harvard University Press, Cambridge, MA), 1991.

[Nas96] J.F. Nash. *Essays on Game Theory*. Edward Elgar, Cheltenham, 1996.

[Ost90] E. Ostrom. *Governing the Commons*. Cambridge, 1990.

[Pia95] J. Piaget. *Sociological Studies*. Routlege, London, 1995.

[Pie91] B.C. Pierce. *Basic Category Theory for Computer Scientists*. MIT Press, 1991.

[PJ04] M. Patton and A. Jøsang. Technologies for trust in electronic commerce. *Electronic Commerce Research Journal*, 4:9–21, 2004.

[Qui98] A. Quinton. *Hume*. Pheonix, 1998.

[Ras01]	E. Rasmusen. *Games and Information (Third edition)*. Blackwell publishing, Oxford, 2001.
[Rec97]	ITU-T Recommendation. X.509 (1997 e): Information technology - open systems interconnection - the directory: Authentication framework. Technical report, 1997.
[Rei65]	F. Reif. *Fundamentals of statistical mechanics*. McGraw-Hill, Singapore, 1965.
[Sca90]	T. Scanlon. Promises and practices. *Philosophy and Public Affairs*, 19(3):199–226, 1990.
[Sch51]	J. Schwinger. Theory of quantized fields i. *Physical Review*, 82:914, 1951.
[Sch53]	J. Schwinger. Theory of quantized fields ii. *Physical Review*, 91:713, 1953.
[Sea69]	J.R. Searle. *Speech Acts*. Cambridge University Press, Cambridge, 1969.
[Sea83]	J.R. Searle. *Intentionality*. Cambridge University Press, Cambridge, 1983.
[She11a]	H. Sheinman. *Promies and Agreements*, chapter Introduction: promises and agreements, pages 3–57. Oxford University Press, 2011.
[She11b]	H. Sheinman, editor. *Promises and Agreements*. Oxford, 2011.
[Sny81]	L. Snyder. Formal models of capability-based protection systems. *IEEE Transactions on Computers*, 30:172, 1981.
[Sto52]	S.J. Stoljar. The ambiguity of promise. *Northwestern University law Review*, 47(1):1–20, 1952.
[Susa]	J. Sussna. Promise theory, devops, and design.
[Susb]	J. Sussna. Promise theory, devops, and design.
[Susc]	J. Sussna. Scaling agile projects using promises.
[Susd]	J. Sussna. Turning requirements into promises.
[SW49]	C.E. Shannon and W. Weaver. *The mathematical theory of communication*. University of Illinois Press, Urbana, 1949.
[Wil12]	J. Willis. Promise theory for dummies. Conference presentation http://www.youtube.com/watch?v=y3yplqTFywY, 2012.

[Win96] I.S. Winkler. The non-technical threat to computing systems. *Computing systems (MIT Press: Cambridge MA)*, **9**:3, 1996.

[Woo02] M. Wooldridge. *An Introduction to MultiAgent Systems*. Wiley, Chichester, 2002.

[WS03] Feng Wan and Munindar P. Singh. Commitments and causality for multiagent design. In *Proceedings of the 2nd International Joint Conference on Autonomous Agents and MultiAgent Systems (AAMAS)*, 2003.

INDEX

$F(C)$, 29
$T(C)$, 29
promise conflict, 96
$\wedge$, 29
$\vee$, 29
employ(), 39
$\alpha_A()$ assessment, 66
μ-promises, 26
def(), 30, 89
$\pm$-promises, 5
$\Rrightarrow$, 29
Valence(), 52
b, 29
Æther, 249

Absorbing an agent, 185
Acceptance promise, 38
Access
    Shared, 154
Accusation, 174
Action of promises, 43
Adjacency, 227, 232
    As information, 227
    Defined, 232
    Matrix, 37
    Promise, 228
Adjacency matrix, 37
Advanced assessment, 69
Advanced boundary condition, 69, 275
Advertising, 171
Affects relation, 231
Agent
    Anonymous, 48
    Boundary, 185
    Ensemble, 77

Equivalence, 73
Inanimate, 14
Interior process, 23
Relativity, 4
Roles, 54
Unspecified, 48
Agents, 2, 26
    Autonomous, 2
    Intermediate, 87, 115
Agents as ensemble, 77
Aggression, 4, 48, 97, 165, 166, 171–174, 191, 273
Agreement, 10
    Definition, 102
    Legal, 226
Alias, 205
Anthropomorphism, 6
Appeal, 166
Application services, 129
Arrow of time, 93, 94
Artificial intelligence, ix
Assembly line, 125
Assessment, 3–5, 22, 65, 69
    Advanced, 69
    Defined, 66
    Retarded, 69
Assimilation of knowledge, 89
Assistance in keeping promises, 79, 182
Assisted promise, 79
Attack, 4, 48, 97, 165, 166, 171–174, 191, 273
Authority, 33, 146, 183, 184
    Brute force, 186
    Calibration, 63
    Central, 187

        Decentralized, 188
        Defined, 184
        Delegation, 187
        Over others, 186
Autonomous agents, 2
Autonomy, 3, 146, 229, 232, 241

Backwards compatibility, 198
Bargaining, 216
Basis set, 229
Becomes relation, 29
Behaviour
        Cooperative, 81
        Intended, 2
        Laws of, 91
        Theory of, 271
Belief, 67
Benefit, 106
Billiard ball, 4
Binding, 69, 78, 225
Biology, 242
Blame, 175
        Reason for, 180
Blueness, 62
Body, 28
Boss, 184
Bottleneck, 130
Boundary, 229, 242
        Agent, 185
        Semantics, 243
Boundary condition, 69,
                92, 239, 273
Boundary conditions, 94
        Advanced and retarded, 275
Brand, 64
Breaking promises, 99
Broadcast, 166
Bundle, 50
        Parameterized, 51

Calibration, 58, 62, 72,
                147, 185, 230, 237
Camouflage, 208
Capacity, 130
Catalyst, 155
Category Theory, 25

Causality, 176, 178, 231
Central authority, 187
Centralization, 3, 146
CFEngine, 250
Circuit switching, 127
Client, 58, 120
Clock synchronization, 71
Close to, 228, 237
Clubs and tenancy, 153
Co-dependent binding, 240
Coalition, 163
Code of conduct, 19
Coercion, 165
Coherence in cooperation, 146
Colour, 62, 227, 228
Command, 165
Commitment, 272
Commitments, 43
Compatibility, 197
Complementarity, 80
Complementarity of $\pm$
                promises, 41
Compliance, 166
Component
        Compatible, 197
        Contentious, 194
        Definition, 193
        Design, 194
        Equivalent, 197
        Interchangeable, 197
        Replacement, 206
Components, 129, 192
        Electronic, 15
Composition of promises, 76
Computer network, 244
Computers, 21
Condition
        Initial or boundary, 92
Conditional promises, 78
Confidence, 107
Conflict, 84
        Give, 97
        Use, 98
Conflicting promises, 96, 194
Conflicts of intent, 10
Conquer, 186

INDEX

Conquering, 185
Consciousness, 6, 7
Consciousness of agents, 23
Consensus, 62
Consensus of knowledge, 91
Consent, 17
Consistency, 82
    Knowledge, 87
Consistency of knowledge, 91
Constitution, 276
Constraint, 28
Containment, 231, 243
Contention, 194
Continuity, 119, 122, 228, 241
Contract
    Agreement, 102
    Definition, 102
    Signing, 103
    Social, 163
Contractor, 168
Cooperation, 9, 40, 81, 211
    Incentives, 225
    Voluntary, 8
Cooperative ensemble, 56
Coordinates, 237
Coordination, 45, 147
Cost of keeping a promise, 107
Cost of organization, 147
Country, 64
Coupling strength, 77
Crystal lattice, 230, 236

Database normalization, 277
Deadlock, 92, 168, 239
Decentralization, 3, 146
Decentralized system, 188
Deception, 24, 45, 208
Decoherence in cooperation, 146
Definition
    Agent knowledge, 86
    Agreement, 102
    Altruist source, 224
    Assimilated knowledge, 89
    Assisted promise, 80
    Authority, 184
    Authority over others, 186

Broken promise, 99
Conditional promise, 78
Consistent knowledge, 87
Contract, 102
Cooperation, 82
Deception, 45
Emergent behaviour, 75
Empty promise, 44
Equivalence under
    observation, 73
Exact and inexact
    promises, 44
Goal, 49
Homogeneous promise
    bundle, 51
Imposition, 32
Incompatible/exclusive
    promises #, 95
Knowable information
    about promises, 30
Move and strategy
    in games, 217
Naming of component
    function, 204
Net valence of promise
    graph, 53
Occupancy, 149
Opportunist sink, 224
Overcommitting, 52
Parameterized promise
    bundle, 51
Policy dilemma, 222
Promise adjacency
    matrix, 37
Promise bundle, 50
Promise conflict, 96
Promise matrix, 37
Promise or μ-Promise,
    27
Promise proposals, 28
Promise refinement, 100
Promise to cooperate, 82
Promise to subordinate, 82
Responsibility, 179
Right to X, 182
Role by appointment, 55

290 INDEX

Role by association, 55
Role by coordination, 56
Scope, 31
Scope of common knowledge, 91
Signing of proposal, 103
Subordination, 82
Tenant host, 151
Tit for tat, 221
Treaty agreement, 103
Two person iterated game, 217
Use-promise saturation, 53
Utilization of agent, 53
Valence of agent, 52
Delegation of authority, 187
Delivery, 115
Delivery chain, 119
Democracy, 163, 186
Dependencies, 202
Dependency conflict, 202
Desired end state, 119
Differential equation, 275
Differentiated behaviour, 57
Dilemma game, 217
Dimensions
    Hidden, 230
Directed invitation, 167
Direction, 228
Disordered state, 245
Dispatcher, 60
Dispatcher role, 60, 130
Disruption, 174
Distance, 227
Distinguishability, 72, 234
Downstream principle, 178
Duality of ± promises, 41
Dynamics, 25, 85, 271

Economics, 25
Economics of organization, 147
Effective action, 77
Effective action of promises, 43
Einstein, Albert, 227
Electronic components, 15

Electrons, 236
Emergent behaviour, 68, 75
End-to-end delivery, 115, 238
End-to-end law, 125
Ensemble, 77
Entanglement, 240
Epidemic
    And trust, 275
Equals relation, 29
Equilibration, 62
Equilibrium, 92
    Nash, 217
equilibrium, 251
Equivalence, 73
Euclidean norm, 274
Even horizon, 233
Event horizon, 242
Event-driven agent, 70
Events
    Defined, 70
Evidence, 67
Evolution of behaviour, 85
Exact promises, 44
Example
    Agent modelling, 138, 192
    Assessment, 66
    Backwards compatibility, 198
    Bundle, 50
    Compatibility, 198
    Component replacement, 206
    Components, 192
    Contention, 195
    Deadlock, 93
    Division of labour, 137
    Electronics, 15
    End-to-end delivery, 127
    Equilibria, 94
    Frame relay, 127
    Imposition, 41
    Imposition vs promise, 41
    Imprecise promises, 96
    Inexact promise, 44
    Inferred promise, 69
    Interchangeability, 198
    Intermediaries, 115

INDEX

Laws of behaviour, 68
Measurement, 71
Non-functional value, 105
Obligation, 41
Observability, 70
Parameterized bundle, 51
Promise conflict, 96
Promise type, 28
Promise vs imposition, 41
River intent, 6
Roles, 55
Transducer, 61
Trust friends, 111
Trust money, 110
Unspecified agents, 48
Value, 24, 105
Verified behaviour, 27
Versioning, 197
Exclusive promises, 95
Expectations, 15, 18, 111
Expiry of promise, 85

Favours, 166
Fire and forget, 119
Fitness for purpose, 208
Flocks of birds, 228
Flooding, 166
Follow a leader, 62
Follows relation, 231
Force, 19, 185
Force majeur, 17
Forced behaviour, 19
Formation of hierarchy, 146
Frame relay, 127
Free speech, 183
Freewill, 6, 7, 23, 274

Gambling, 111
Game, 209
    Dilemma, 217
Game Theory, 92, 165, 209, 277
Game theory, 25
Games
    Bargaining, 216
    Constant sum, 213
Gas phase, 245

Generalizes relation, 231
Global consistency, 76, 82, 87
Global namespace, 204
Goal, 9
Goals, 49
Godlike observer, 228, 238
Gradient, 234
Granted
    Taking something for, 7
Graph, 127
Graphs, 36, 57

Handshake, 78, 225
Help desk, 14
Hidden dimensions, 230
Hierarchy, 146
    Economics of, 147
    Emergence of, 148
Human rights, 181
Human systems, 163
Humanity, 6
Humans, 2
Hume, David, 271

Idempotence of promises, 47
Ideology, 3, 146
Imposing, 1
Imposition, 4, 17, 120, 185
    Accepting, 40
    And law, 226
    And Obligation, 19, 33
    And push, 121
    And spacetime, 231
    Breaking deadlock, 92
    Client, 58
    Defined, 32
    Dispatch, 60, 130
    Ephemeral, 120
    In processes, 116
    Inequivalence with
        use promise, 40
    Kinds of promises, 33
    Obligation, 19
    Of order, 234
    Polarity, 39
    Pre-existing promises, 4

Relation to promise, 33
Relative, 119
Requirement, 201
Self, 19
Subordination, 84
Top down, 118
Transfer of intent, 38
Workflow, 120
Inanimate agents, 14
Incentives, 225
Incompatible promises, 95
Incomplete information, 97
Indirection, 100
Indistinguishability, 73
Inducing behaviour, 17, 19, 21, 33
Inferred promise, 68
Information, 2
    Integrity, 87
    Technology, 8, 21, 127, 129, 161, 204, 250
    Transfer, 274
    Types, 7
Initial condition, 92
Institutions, 137
Integration
    Of components, 207
Integrity, 87
Intent, 271
    And freewill, 7
    As a possible choice, 7
Intention, 4, 16
    Generalized, 16
    Hidden, 24
    Possible, 24
    Transfer, 38
Intentionality, 4, 16
Interaction, 69
Interchangeability, 197
Interfaces, 199
Intermediaries, 115
Intermediate agent law, 87
Intermediate agents, 87
Intrinsic properties, 230
Intrusion, 165
Invariant, 274

Invariants, 13
Invitation, 165
    As imposition or promise, 169
    Directed, 167, 170
    Layers of, 172
    Open, 165, 167

Judgement, 174

Kinds of promises, 33
Knowledge, 3, 30, 86, 232
    assimilation, 89
    Consensus, 91

Lattice, 233
Law, 226
    Assisted promise, 79
    Assisted promise (exact), 79
    Composition, 76
    Composition of promises, 76
    Conditional promise, 78
    Conflicting promises, 84
    Quenching of assisted promise, 79
Leader
    Follow, 62
Leadership, 190
Legal profession, 271
Lemma
    Adjacency is tenancy, 157
    Causation partially ordered by dependency, 152
    Cooperative ensemble, 56
    Disruption implies dependency, 174
    Games are bilateral promises, 219
    Inequivalence of $-$ promise with $+$ imposition, 40
    Irreducible promises are conditional, 145
    Only positive information in promise, 68

Promise notation is idempotent, 47
Tenancy flows in direction of used resource, 151
Liability, 175
Lie, 24
Lies, 45, 208
Life-cycle of promises, 85
Light bulb, 2
Litter
    Don't drop, 169
Load balancing, 130
Local minima, 146
Locality, 232
Logistic chain, 122
Logistics, 115
Long range order, 229, 233, 277

Mach's principle, 238
Macrostates, 274
Management, 190
Manager, 184
Mandate, 40, 185, 186, 190
Many worlds, 2, 4, 30, 71, 85, 105, 206, 273
Market forces, 146
Marketing, 105
Markov chain, 235
Matroid construction, 62, 229
Mean Time
    Before Failure, 106
    To Keep a Promise, 106
    To Repair, 106
Measurement, 65, 69
    Algebra, 71
Mechanics, 76
Mechanisms and Promises, 1
Membership, 153, 228
Memory
    And scalar promises, 230
Message driven agent, 70
Metric space, 237
Micropromises, 26
Microservices, 207
Modularity, 206
Molecule, 241

Moral assessment, 178
Morality, 4, 8, 271, 272
Motion, 245
    First kind, 246
    Second kind, 248
    Third kind, 249
MTBF, 106
MTTR, 106
Multi-agent systems, 272
Multi-component architecture, 129
Multi-tenancy, 154

Naming of agents, 204
Nash equilibrium, 92, 217
Neighbour, 228
Network
    Computer, 127, 244
Newton's laws, 91
Newton, Isaac, 227
Non-cooperative Game Theory, 211
Non-local order, 233
Norms, 19
North
    Concept of, 228
Nyquist-Shannon sampling, 23

Obligation, 1, 4, 19
    Against, 20
    And imposition, 19, 33
    And rights, 182
    Defined, 33
    Example, 41
    In favour of, 20
    Meaning, 19
    Promissory, 8
Observability, 23
Observable, 229
Observation, 16, 65
Observer, 2
    Godlike, 228, 238
Occupancy, 148, 227
    Scaling, 158
Offence/Offense, 174
Open invitation, 165, 167

Optimization, 146
Orchestration, 10
Order
    Civil, 186, 190, 226
    Of agents, 228
Order relation, 235
Organization, 135, 137
    Cost of, 147
OSI model, 152
Outcome, 66
    Assured, 190
Overlap, 234, 244
Overriding, 100
Ownership, 148

Parallel organization, 136
Partial order, 237
Patterns of promises, 50
Payoff matrix, 277
Periodic table, 237
Permission, 181, 183
Petri net, 251
PGP, 187
Phase transition, 245
Physics, 1, 2, 23, 25,
    145, 149, 164, 168, 173, 271,

Pipeline, 115
    Delivery, 119
Pixels, 64
Polarity
    Impositions, 39
    Promises, 38
Polarity of promise, 5
Politics, 181
Precedes relation, 231
Preorder relation, 240
Pretty Good Privacy, 187
Principle
    Downstream, 178
Prisoner's dilemma, 212
Privilege, 154, 181, 183
    And rank, 182
Probability, 111, 210
    Bayesian, 108
Processes

And Promises, 23
Production line, 125
Projection theorem, 36
Promise, 4
    $\pm$, 5
    About, 46
    Absolute, 119
    Acceptance, 38
    Action, 43
    And obligation, 18
    Binding, 69, 78, 225
    Body, 28
    Broken, 99
    Bundle, 50
    Composition, 76
    Conditional and tenancy,
        156
    Conflict, 84, 96, 194
    Constraint, 28
    Continuity, 122
    Coordination, 45
    Deception, 24, 45
    Definition, 27
    Effective action, 43
    Elementary, 36
    Emergent behaviour, 68, 75
    Empty, 44
    Exact, 44
    Exclusive, 95
    Expiry, 85
    Formalizing, 26
    Graph, 36
    Heuristic, 13
    History of concept, 8
    How to use, 9
    Idempotence, 47
    Implicit, 17
    Inanimate, 14
    Incompatible, 95
    Inexact, 44
    Inferred, 68
    Kept, 65
    Kinds, 33
    Language, 255
    Lie, 24
    Lies, 45

INDEX 295

    Life-cycle, 85
    Literature about, 18
    Meaning, 1
    Model, 15
    Motivation for, 13
    Not kept, 65
    Notation, 29
    Outcome, 66
    Patterns, 50
    Polarity, 5, 38
    Proposal, 28
    Quenching, 79
    Responsibility, 175
    Scalar, 63, 139, 149, 230, 232
    Systemic, 131, 135, 193
    Tenets, 3
    Tensor, 149
    To use, 38
    Unspecific, 200
    Value, 23, 104, 105
    Vector, 149
    Versus obligation, 18
Promise proxy, 14
Promise types, 47
Promisee, 16, 26
Promiser, 14, 26, 272
Promising, 1
Promisor, 14, 272
Promissory obligation, 8
Propaganda, 171
Proposal, 28
Provenance, 176
Proximity, 232
Proxy, 115, 122
Proxy promise, 14
Public key, 204
Publishing of intent, 165
Pull vs push, 41
Purchase order, 168
Purpose, 271
Push model, 121
Push vs pull, 41, 80, 119, 120

Quality of outcome, 107
Quantum Mechanics, 277

Quenching, 79

Random variables, 67
Rank
    And Privilege, 182
Ranking of promises, 100, 107
Rational decisions, 209
Reasoning, 95
Reducing uncertainty, 9
Refinement of promises, 100
Regression test, 199
Relation
    Affects, 231
    Follows, 231
    Generalizes, 231
    Order, 235
    Precedes, 231
    Preorder, 240
    Special case of, 231
Relative imposition, 119
Relativity, 4, 10, 30, 71, 85, 105, 206, 273
Reliability, 107
Reprimand, 175
Reputation, 112
Requirements, 9, 201
    Minimum incentive, 225
Responsibility, 175, 190
    Assuming, 180
    Conditional promises, 176
    Defined, 179
Retarded assessment, 69
Retarded boundary condition, 69, 120, 275
Reusability, 137, 196
Reversibility, 93, 277
RGB promise, 64
Right
    To impose, 186
Rights, 181
    Bill of, 276
    Definition, 182
    Demanding, 183
    Free speech, 183
    I know my..., 183
    Seeking, 183

Risk, 107
River example, 6
Robust system, 133
Roles, 54, 138, 274
    By appointment, 55, 187
    By association, 55
    By cooperation, 56
    Client, 58
    Consumer, 58
    Graphical interpretation, 57
    Matroids, 62
    Peer, 59
    Queue, 60
    Server, 59
    Server pool, 59
    Transducer, 61
Root Cause Analysis, 178
Rule
    Absolute promises
        versus relative impositions, 119
    Advertise exclusivity, 99
    Component dependency
        can be made version specific, 202
    Exclusive promises
        should be unique, 98
    Idempotence of promises, 47
    Limit scope of exclusive
        promises, 99
    Promise Continuity, 122
    Reusable components
        and avoidance of exclusive prom 203
    Separate events have
        separate types, 47
    Simultaneous promises
        must have different types, 99
    Uniqueness of naming
        ± promises, 206

Safety, 107
Sampling of information, 70
Scalability, 129
Scalar promise, 63, 139, 149, 230
    Defined, 230
    Singular, 232
Scope, 3, 30, 31, 68, 91
    Obligations, 19
Security, 107, 178, 181, 187, 276
Self-organization, 245
Semantic spacetime, 227
Semantics, 2, 25, 78
Semi-permeable membrane, 242
Separation of concerns, 137
Serial organization, 136
Server, 59, 121
Server pool, 130
Service
    Negotiation, 168
Service view, 80
Services, 129
Shared resource, 154
Shoals of fish, 228
Signing
    Contract, 103
Single point of failure, 134, 156, 177, 224
Social contract, 163
Societies and Promises, 1
Society, 181
Space, 227
    Metric, 237
Spam, 171
Spanning set, 229
Spanning tree, 273
Special case of, 231
Specialization, 137
Standardized promises, 62
Statecraft, 186, 190, 226
Strategies, 215
Strong coupling, 134
Subordination, 82, 83, 92, 186, 188
    Calibration, 63
Superagent
    And long range order, 241
Swarm intelligence, 75
Swarms, 228

INDEX

Symbiosis, 92, 148, 153, 159, 186, 187
Symmetry
    Behavioural, 57
    Deadlock, 92
    Time reversal, 42
Symmetry breaking, 94, 245, 273, 275
System, 50, 135
System integration, 207
System promises, 193
Systems
    Human, 163

Takeover, 186
Taking for granted, 7
Taxonomy of roles, 273
Tenancy, 148
    Asymmetric, 157
    Conditional promise, 156
    Forms of, 153
    Laws of, 150
    Multi-, 154
    Remote, 156
    Scaling, 158
    Semantics of, 150
Tensor promise, 149
Tensor promises, 231
Territory, 154
Theorem
    Consistent knowledge, 87
    Constant sum game, 213
    Projection for promises of n-th kind, 36
Thermodynamics, 274
Third party observer, 232
Time, 227
    To Keep a Promise, 106
Time reversal symmetry, 42, 93
Time to deliver, 107
Time to value, 107
Timescales, 234, 241
Tit for tat, 221
TLS, 187
Tools, 271
Top down, 118

Topology, 231
Trade, 216
Transducer, 61
Transfer of intent, 38
Transformation chain, 125
Transitivity, 231
Transport network, 127
Transport view, 80
Treaty, 103
Trust, 8, 107
    And epidemic, 275
    And non-locality, 238
    And promises, 23
    Defined, 108
Trust architecture, 187
Trusted Third Party, 63, 187
Types of information, 7

UML, 251
Uncertainty, 10, 18, 91
Undifferentiated behaviour, 57
Unexpected behaviour, 4
Universality, 163, 182
Usage
    Naming, 205
Use promise, 38
Utility matrix, 277

Valency, 52, 154
Value of a promise, 5, 104
Value of promises, 23
Vector promise, 149
Vector promises, 231
Vector space, 233
Velocity
    Non-local nature, 248
Versioning, 197, 202
Virtual circuits, 127
Virtualization of promises, 156
Voluntary cooperation, 8, 209
Voting, 163
Vulnerability, 107

Web
    Of trust, 187
Web of trust, 188

Workflow, 115
    Imposition, 120
Worlds, 2, 4, 30, 71, 85, 105

X.509, 187

Zero-sum games, 213

INDEX

Acknowledgements

We have been working sporadically on Promise Theory since 2004. During that time we have talked with many individuals who have helped to shape our thinking.

Jan: I am grateful to Kees Middelburg and Bas van Vlijmen for many conversations about philosophical aspects of informatics including promises. Thanks also to Inge Bethke for collaboration on [BBB07].

Mark: For the first four years, I worked with PhD student Siri Fagernes [BF07a, BF07b, BFa, BFb, BF06] to explore many applications of the basic ideas. Alva Couch also helped to develop some of the early applications [BC06]. For the second edition, I'd like to thank Daniel Mezick for valuable conversations and deep discussions on the application of promise ideas in social contexts.

I would like thank Google for hosting the 2008 talk that started the ball rolling for many practitioners in Information Technology[Bura]. John Willis' brave belief in the power of promises and CFEngine resulted in an excellent talk of his own[Wil12]. Jess Sussna has popularized the principles in a series of articles[Susd, Susc, Susa, Susb], and more recently Paul Borrill and Mike Dvorkin have applied promise theory to their own areas of IT.

About the Authors

Jan Bergstra is a retired Dutch computer scientist, living in Utrecht. He has worked at the Institute of Applied Mathematics and Computer Science of the University of Leiden, and the Centrum Wiskunde & Informatica (CWI) in Amsterdam. In 1985 he became Professor of Programming and Software Engineering at the Informatics Institute of the University of Amsterdam and Professor of Applied Logic at Utrecht University. His work has focused on logic and the theoretical foundations of software engineering, especially on formal methods for system design. He retired as a full professor in the end of 2016. He is best known for work on algebraic methods for the specification of data and computational processes in general. Jan's current affiliation is Minstroom Research BV, and he can be reached via `janaldertb@gmail.com`.

Mark Burgess is a British theoretical physicist, turned computer scientist, living in Oslo, Norway. After authoring and consulting for the IT industry and holding a number of research and teaching positions, he was appointed Professor of Network and System Administration at Oslo University College in 2005, which he held until resigning in 2011 to found the CFEngine company. He is the originator of the globally used CFEngine software as well as founder of CFEngine AS, Inc. He is the author of many books and scientific publications, and is a frequent speaker at international events. Mark Burgess may be found at `www.markburgess.org`, and on Twitter under the name `@markburgess_osl`.

Made in the USA
Coppell, TX
15 March 2020

16882534R00185